THE BIG HOP

The North Atlantic Air Race

GAVIN WILL

BOULDER
PUBLICATIONS

Library and Archives Canada Cataloguing in Publication

Will, Gavin, 1961-
The Big Hop: The North Atlantic Air Race / Gavin Will.

Includes bibliographical references and index.
ISBN 978-0-9730271-8-1

1. Aeronautics--Newfoundland and Labrador--History.
2. Transatlantic flights--History. I. Title.

TL523.W448 2008 629.13'09718 C2004-906023-6

Published by Boulder Publications
Portugal Cove-St. Philip's, Newfoundland and Labrador, Canada

www.boulderpublications.ca

Book design and layout by John Andrews Printed in Canada

We acknowledge the financial support of the Government of Newfoundland and Labrador through the Department of Tourism, Culture and Recreation.

Newfoundland Labrador

CONTENTS

INTRODUCTION

This book was inspired by Robert Tait, who wrote an unpublished manuscript on the exploits of aviators who used Newfoundland as their departure point for reaching Europe after the First World War.

Tait, a war veteran who became Newfoundland's consul in Boston, was among the throngs who gathered on fields in St. John's and Harbour Grace to watch as mechanics and pilots prepared their aircraft for journeys across the ocean. Among the historic flights he witnessed was the first successful Atlantic crossing by John Alcock and Arthur Whitten Brown in 1919.

Today, in an era where passengers travel in temperature-controlled jumbo jets that cruise serenely over continents and oceans, it is difficult to appreciate the dangers that early aviators faced as they strove for glory.

By 1919, powered flight was barely beyond its infancy; the Wright brothers had flown their first rickety plane only sixteen years earlier, in 1903. Significant improvements in aircraft performance soon followed, particularly during the First World War when they were used for reconnaissance and "dogfights" with enemy planes. Bombers, which made an appearance by the end of the war, were converted for long-range travel after the conflict ended; several of them were used in transatlantic attempts in 1919.

Despite steady improvements in technology, prior to the mid-1930s flying was hazardous – particularly in inclement weather – and aircraft engines were prone to failure. Reliable aerial navigational techniques were primitive, and radios even worse. But early aviators knowingly risked their lives to cross the world's oceans because the rewards were tantalizing: ticker-tape parades attended by hundreds of thousands of people in New York and audiences with European royalty were virtually guaranteed; and prize money was offered by newspapers and wealthy patrons.

During this brief era pilots shared the public spotlight with movie stars; continental air races, endurance flights, daredevil antics, altitude records, and ocean crossings fed an insatiable public interest. Newspapers, magazines, radio stations, and cinematic newsreels provided blanket coverage of pilots and their feats during the 1920s and 1930s.

But the public was not content to watch its heroes from afar – people yearned for the day they themselves would be able to fly. Crossing oceans by air was particularly enticing, because planes offered a much quicker alternative to passenger ships.

Until the mid-1930s passenger flights across the Atlantic were conducted by airships, most notably the *Graf Zeppelin* and the ill-fated *Hindenburg*. But the future lay with fixed-wing passenger planes, and by the mid-1930s several large airline companies and aircraft manufacturers had emerged that were capable of developing reliable, cost-efficient aircraft.

The pioneering era of aviation ended with the onset of the Second World War. Technological advancements made during the war years, combined with the proliferation of modern airports with paved runways, made passenger flights safe and convenient.

This book chronicles the early years of transatlantic flight, with particular attention being paid to Newfoundland's crucial – but little recognized – role in creating the world of aviation we recognize today.

In writing this account I am indebted to numerous people, all of whom volunteered their knowledge and advice. They include aviation enthusiasts Darrell Hillier and Gary Hebbard, who freely offered their expertise, editor Iona Bulgin, staff at the North Atlantic Aviation Museum in Gander, and the town of Botwood. I am particularly grateful to graphic designer John Andrews, whose skills and professionalism have made this book an object of beauty; and to my father, Raymond Will, who reviewed this manuscript at its earliest stages and had the courage to tell me the truth.

CHAPTER 1

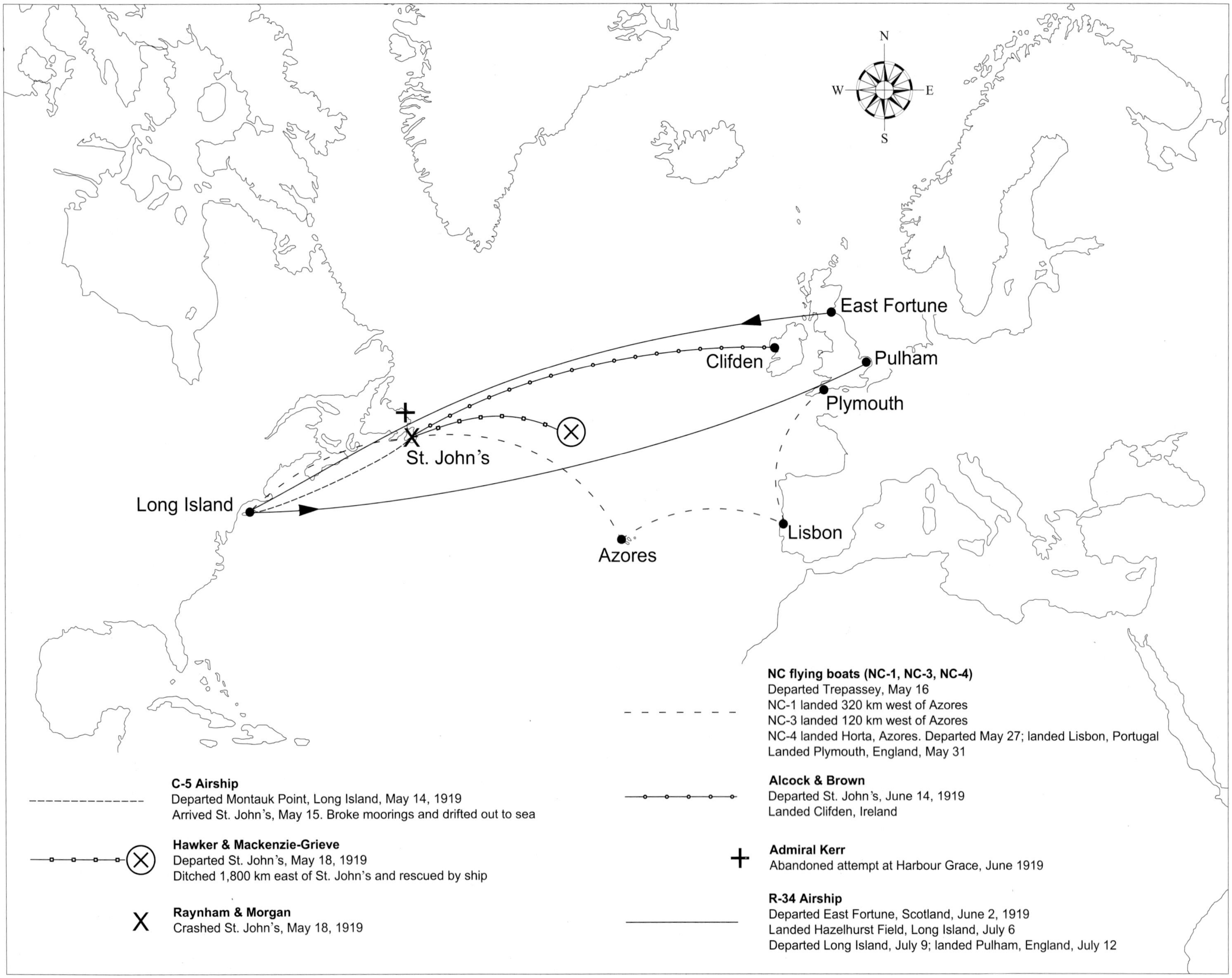
N
W
E
S
East Fortune
Clifden
Pulham
Plymouth
St. John's
Long Island
Azores
Lisbon
NC flying boats (NC-1, NC-3, NC-4)
Departed Trepassey, May 16
NC-1 landed 320 km west of Azores
NC-3 landed 120 km west of Azores
NC-4 landed Horta, Azores. Departed May 27; landed Lisbon, Portugal
Landed Plymouth, England, May 31
C-5 Airship
Departed Montauk Point, Long Island, May 14, 1919
Arrived St. John's, May 15. Broke moorings and drifted out to sea
Alcock & Brown
Departed St. John's, June 14, 1919
Landed Clifden, Ireland
Hawker & Mackenzie-Grieve
Departed St. John's, May 18, 1919
Ditched 1,800 km east of St. John's and rescued by ship
Admiral Kerr
Abandoned attempt at Harbour Grace, June 1919
Raynham & Morgan
Crashed St. John's, May 18, 1919
R-34 Airship
Departed East Fortune, Scotland, June 2, 1919
Landed Hazelhurst Field, Long Island, July 6
Departed Long Island, July 9; landed Pulham, England, July 12

RACING THE ATLANTIC

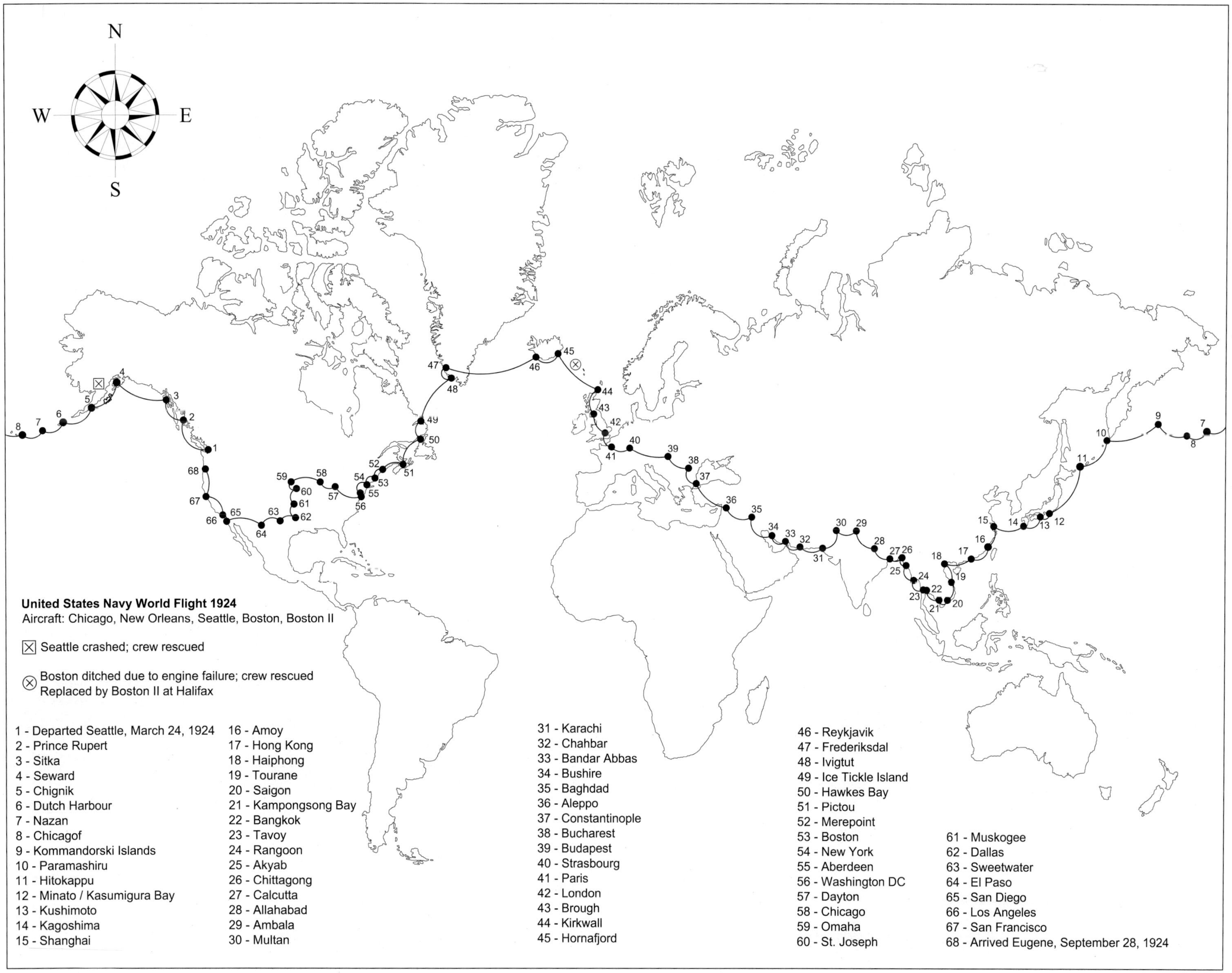

In February 1919, a passenger liner glided through the narrow waterway that cuts between the grey austere cliffs guarding the entrance to St. John's, the capital city of Newfoundland.

The voyage from England had been difficult for those on board the ship, and not just because of the notoriously stormy weather encountered in the mid-Atlantic: passengers and crew had faced an outbreak of influenza, a disease that was proving as deadly as the bullets that had killed millions of men during the four years of conflict that had ended just three months earlier.

Among those who contracted influenza during the voyage was Charles Morgan, a veteran British aviator now working for the Martinsyde aircraft manufacturing company. Morgan, who had recovered by the time the ship reached port, immediately began searching for an open field suitable for launching an aircraft across the vast expanse of the Atlantic Ocean.

Morgan's hunt for a makeshift airfield in St. John's started a whirlwind of speculation, since his activities marked the first tangible evidence

The Atlantic air race of 1919 offered a vision of hope to nations traumatized from the carnage of the First World War. In Newfoundland, residents of St. John's welcomed three British teams, including the team of John Alcock and Arthur Whitten Brown (left). One of the planes, the *Raymor*, crashed during takeoff. The Cochrane Hotel in St. John's became a temporary home to aircrews competing for the £10,000 *Daily Mail* transatlantic prize. From left: Kenneth Mackenzie-Grieve, Frederick Raynham, Harry Hawker, and Charles Morgan.

Close-up view of the nose of Hawker and Mackenzie-Grieve's Sopwith aircraft, after being recovered by USS *Lake Charlotteville*. The wreckage and a mail bag were salvaged and taken to Falmouth, England.

that an air race across the Atlantic was set to begin.

The focus for this competition was provided by Lord Northcliffe, owner of Britain's *Daily Mail* newspaper, who in 1913 offered a prize of £10,000 (US $50,000) to the first aircraft to fly between Europe and North America.

Because of the outbreak of war, no serious plans to claim the prize were made prior to 1918. By the latter stage of the conflict, however, dramatic improvements had been made to aircraft range and reliability, due largely to the development of bombers capable of reaching cities and factories deep inside enemy territory.

In 1919, as companies and airmen from several countries publicly mulled over their intentions to compete for the prize, Northcliffe altered the rules to favour British flyers. Competitors from former enemy nations were now banned outright, a non-stop stipulation was added, and claimants had to land in either Great Britain or Ireland.

The non-stop amendment was

made to exclude the US Navy, which was planning to fly specially designed Navy Curtiss flying boats to the Azores from Newfoundland and then continue on to Lisbon and England.

With the rules thus altered, Britain's largest aircraft manufacturers assembled teams of flyers and mechanics. As an advance man for the Martinsyde team, Morgan spent several frustrating days in St. John's tramping through snowdrifts in search of sufficient flat land to launch an aircraft.

"I had the pick of what there was, but after spending days going all over the place and plodding through the snow with a kind of alpenstock [staff], the best I could find was a field beside Quidi

Harry Hawker (far right) and Kenneth Mackenzie-Grieve became the first airmen to attempt a non-stop transatlantic flight. They took off from Mount Pearl on May 1, 1919, but engine trouble forced them to ditch the aircraft into the ocean. Following their rescue, the *Daily Mail* awarded them a consolation prize of £5,000.

The *NC-1*, at Trepassey (above), was one of three United States Navy flying boats to make a transatlantic attempt in May 1919. Trepassey was selected as the launching point because of its sheltered harbour. The *NC-4* (right) was the only one to reach the Azores intact. It then flew to Portugal and Great Britain.

Vidi Lake just outside St. John's," Morgan wrote. "I caught influenza on the way over, which depressed me, and the disappointment over the aerodrome turned me really gloomy."

Morgan may have harboured misgivings about his pending adventure, but residents of St. John's were eagerly anticipating the competition. Few Newfoundlanders, aside from soldiers returning from Europe, had ever seen an aircraft, and there is no record of anybody from this small, independent nation flying as either a pilot or a passenger prior to 1919. But now, even as people mourned the carnage wrought by the First World War, excitement was building as Newfoundland prepared to host the greatest air race the world had ever witnessed.

Having found an airstrip for Martinsyde, Morgan went back to England, but he returned to St. John's in mid-April with pilot Frederick Raynham and a biplane, the *Raymor*.

By then, St. John's was already hosting another British team, this one sponsored by the Sopwith Aviation Company, which was led by pilot Harry Hawker and navigator Kenneth Mackenzie-Grieve. With Quidi Vidi in St. John's already taken, the Sopwith crew sought another airstrip for their biplane, and settled on Glendenning's farm on Brookfield Road in what is now the city of Mount Pearl. After several weeks spent reassembling and testing their aircraft, the Martinsyde and Sopwith teams came to a gentleman's agreement whereby the first team ready to take off for Europe would alert the other prior to their departure.

Not only were the British teams in a race with each another, they were facing competition from the US Navy. Although their flying boats were ineligible to claim the *Daily Mail* prize because of a planned stop in the Azores, the Americans were intent on winning a partial victory over their British competitors. In its bid to cross the Atlantic the navy ordered four custom-built flying boats from Curtiss Aeroplane and Motor Company, then the dominant aircraft manufacturer in the US. Dubbed NCs, or "Nancies," three of the planes took part in the expedition, while the fourth was used for spare parts.

On May 10, two of the flying boats, the *NC-1* and the *NC-3*, flew to the Newfoundland community of Trepassey. A few days later they were followed by the *NC-4*. Located 150 kilometres from St. John's at the southeast tip of the Avalon Peninsula, Trepassey became a temporary US Navy base, with ships anchored offshore in a support role for the aircraft.

The *Daily Mail* contestants had no direct support from the British military. However, the Trepassey-based flying boats were supported by the full might of the US fleet, with 41 navy ships positioned along the route to the Azores to serve as beacons to the aircraft. The Americans, led by Commander Richard Byrd, intended to win bragging rights for achieving the first non-stop flight across the ocean, even if the country did not have a plane capable of travelling that far without stopping to refuel.

Without Byrd's leadership and persistence, it is doubtful if the US would have been as keen to embark upon this ambitious transatlantic venture. Byrd, who had joined the navy in order to become a pilot, had hoped to take part in combat during the First World War. To his disappointment an incurable leg injury kept him out of the European war, and he resigned in frustration from the navy in 1916. Byrd re-

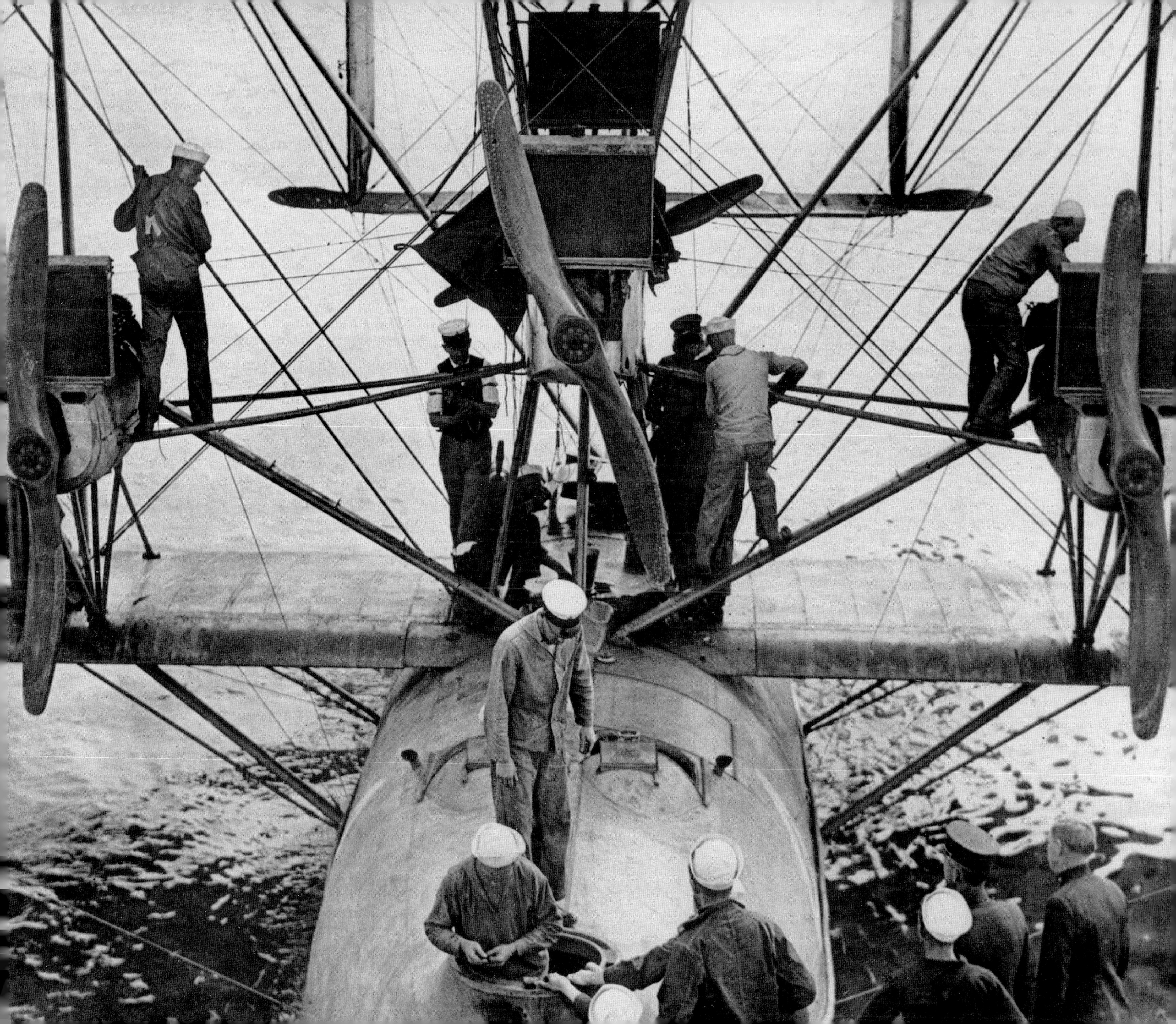

mained active as a reservist, however, and when the US joined the war in 1917 he again hoped to fly a plane across the Atlantic and join American forces fighting in Europe.

Instead, in mid-1918 Byrd was appointed to establish a US Navy station in Nova Scotia, at Baker's Point on the Dartmouth side of Halifax harbour. Baker's Point, which is now part of Canadian Forces Base Shearwater, became home to several flying boats, airships, and kite-balloons, which were used to guard Allied ships from German U-boats. A second American surveillance base was established at North Sydney, Nova Scotia. Another, which was planned for Cape Broyle, a community situated halfway between St. John's and Trepassey, was cancelled when the war ended in November 1918.

With peace, the navy focused on Byrd's transatlantic project, using its new NC flying boats. Byrd had hoped to fly one of the aircraft across the Atlantic, but again his ambitions were dashed, this time by an edict issued by his superiors that the planes be crewed entirely by men who had not participated in a combat zone. The rationale for this rule remains obscure, but it applied to Byrd.

The navy did allow Byrd to command the American transatlantic effort, however, which included a dramatic element other than the flying boats.

On May 15, 1919, spectators dashed to the shores of Quidi Vidi Lake and watched in awe as a blue cigar-shaped airship sailed serenely over the city. The US Navy planned to fly the *C-5* blimp non-stop to Europe from Newfoundland. Under the command of Emory Coil, it had flown from Montauk Point, Long Island, and, after arriving in St. John's, was now tethered to the ground close to the Martinsyde biplane.

The *C-5* was awaiting the arrival of Byrd, who was travelling by train to St. John's from Trepassey to meet the airship, before Coil and his men set off across the ocean. But the crew underestimated the ferocity of maritime storms in Newfoundland, and high winds suddenly tore the *C-5* from its moorings on the day of its arrival in St. John's. Two crewmen leaped to the ground – one of them sustaining serious injuries – before the airship was carried out to sea. It was never seen again.

Although the airship was out of the contest, the Americans still had their flying boats, which were now ready to attempt an ocean crossing. With the two British air teams bogged down in testing their machines, many of the dozens of reporters covering the air race boarded a train to visit the windswept, barren landscape of Trepassey. They would not be disappointed, for on May 16 they watched as the three flying boats took off and headed southeast for the Azores. The journey would prove to be a perilous one, even with support from the US fleet. Each flying boat carried a radio, but all three instruments proved unreliable. As a result, the crews had to rely heavily upon visual beacons provided by the navy's ships along the route.

A dense fog bank soon rendered this assistance useless and the three flying boats flew blind for much of the journey. The *NC-1*, under Lieutenant Commander Peter Bellinger, encountered engine trouble and landed at sea 320 kilometres from the Azores. A ship rescued the

The *C-5* blimp, seen here shortly after it arrived in St. John's in May 1919. High winds tore the blimp from its moorings and carried it out to sea. The *C-5* was one of a series of lighter-than-air craft designed by the US Navy following the First World War. The crew of six men flew 1,650 kilometres from Long Island to St. John's in 25 hours and 50 minutes, on the first leg of a planned transatlantic journey.

crew, but the plane sank. The *NC-3*, piloted by Commander John Towers, became lost in the fog and also landed on the ocean, 120 kilometres short of the Azores. Rough seas damaged the *NC-3* to such an extent that it could not take off again, but Towers and his crew were able to guide the craft by sea to Ponta Delgado in the Azores using the plane's three engines.

Only the *NC-4*, under the leadership of Lieutenant Commander Albert Read, succeeded in flying to the town of Horta in the Azores. Poor weather and mechanical problems delayed its departure to the Portuguese mainland until May 27. The *NC-4* landed in Lisbon after a ten-hour flight and reached its final destination, Plymouth, England, on May 31.

The departure of the flying boats from Trepassey accelerated Hawker and Mackenzie-Grieve into action, and they hastily decided to attempt to reach Ireland before the Americans succeeded in reaching Lisbon. On May 18, the Sopwith plane, *Atlantic*, took off and began its 3,000-kilometre flight to Ireland. The plane's undercarriage, which had been intentionally dropped into the ocean near the Newfoundland coast, was salvaged and later placed on display in a museum.

The Sopwith team's departure panicked the Martinsyde team, and within two hours Raynham and Morgan were ready to attempt to catch and overtake the *Atlantic*. Their aircraft, *Raymor*, was heavily weighed down by fuel, however, and upon takeoff it rose only six metres into the air before crashing nose-first into the ground. Morgan, who had already endured hardships from this adventure, suffered severe lacerations to his face. Raynham later made a second attempt, using a different navigator, but the *Raymor* crashed once again. With this failure, Martinsyde ordered its crew to pack up the wrecked plane and return to Britain.

Martinsyde had been afforded this second chance because Hawker and Mackenzie-Grieve had failed to arrive in Ireland and, after several days, had been given up for dead. Memorial ceremonies planned in their honour had to be hastily cancelled one week later when a Danish cargo ship, the *Mary*, arrived in Thurso, Scotland, with Hawker and Mackenzie-Grieve on board. The *Mary*, which did not carry a radio, had rescued the aviators 1,800 kilometres east of Newfoundland. Hawker explained to newspaper reporters afterwards that the engine began overheating soon after takeoff, causing its radiator to boil over. The plane zigzagged for nearly three hours over the main shipping lanes before sighting the Danish ship. Hawker and Mackenzie-Grieve brought their aircraft down at sea, and they were rescued.

"We only wanted a few gallons of water, and there was the whole damned Atlantic underneath us," Hawker said. "That's what made me puke – knowing that we had enough petrol to reach Ireland and not being able to do it because of the water."

Lord Northcliffe gave Hawker and Mackenzie-Grieve a £5,000 consolation prize, but the title for making the first non-stop transatlantic flight remained unclaimed. However, the failure of both the Sopwith and Martinsyde teams gave a pair of latecomers an opportunity to claim the transatlantic title. On May 26, a team from the Vickers aircraft firm had arrived in St. John's, while a group led by British naval officer Admiral Mark Kerr began assembling a massive Handley Page biplane in Harbour Grace.

Finding a makeshift airfield in St. John's proved difficult for the Vickers crew because Martinsyde and Sopwith had already leased the best available land. An agreement was eventually reached to use a farm near Mundy Pond called Lester's Field. Martinsyde did allow

The presence of a four-engine Berlin bomber in the community of Harbour Grace generated considerable excitement. Admiral Kerr and his crew thrilled people across eastern Newfoundland by taking the massive aircraft on test flights throughout the region.

Vickers to uncrate and rebuild their modified Vimy bomber at Quidi Vidi, however, and on June 9 the plane was flown to Lester's Field.

During this time, Vickers' chief test pilot John Alcock and navigator Arthur Whitten Brown, whose parents were American, prepared themselves for the perilous journey ahead. The two men shared similar experiences during the war: they had both been shot down and then served the duration of the war as prisoners, Alcock in Turkey and Brown in Germany.

Meanwhile, Harbour Grace buzzed with anticipation as test flights of the four-engine Handley Page V/1500 Berlin bomber, the largest aircraft built by the Allied forces during the war, began. The Vickers Vimy and Handley Page bombers were similar in that they had been built and tested in the later months of the war, and neither aircraft was sent into action prior to the armistice.

Kerr and his crew of three flew the huge biplane over St. John's in early June, thereby adding an element of urgency to the Vickers team. However, persistent problems with the engine's cooling system forced Kerr to delay his transatlantic bid. This gave Alcock and Brown sufficient

John Alcock (left) and Arthur Whitten Brown took off from Lester's Field (above) in St. John's on June 14, 1919, and made history by making the first non-stop flight across the Atlantic Ocean.

time to complete preparations of their aircraft, which carried a radio and a second fuel tank which was stowed in the plane's bomb compartment.

As the weather warmed, a festive atmosphere pervaded St. John's in anticipation of the Vimy flight. The huge white biplane drew hundreds of people, first to Quidi Vidi Lake and then to Mundy Pond – families brought picnic baskets and small children ran excitedly around the aircraft.

Poor weather over the Atlantic kept the Vimy grounded for several days, but on June 14 conditions were judged safe, and, in the mid-afternoon, Alcock and Brown climbed into the open cockpit and took off for Ireland.

Good fortune accompanied them for the first hour as the weather remained clear. Their luck soon changed when the small propeller driving the wind-driven electrical generator which supplied power to the radio snapped off. They were now completely alone with no contact with the outside world at all.

Weather conditions then turned against them as they passed through dense clouds and fog. Unlike pilots of today who use electronic navigational aids, early aviators had to rely upon the sun and stars to guide their planes over the open ocean. With no radio and unable to see anything outside his plane a pilot would be flying blind.

During the night the Vimy entered dense cloud and Brown could no longer navigate the plane. He and Alcock suddenly had no idea which way was up, or whether they were flying parallel with the ocean. When the pitch of the engines began increasing in volume, they realized that their plane was plunging towards the Atlantic but had no idea which way to turn.

"Apart from the changing levels marked by the aneroid [an air pressure gauge], only the fact that our bodies were pressed tightly against the seats indicated that we were falling," Brown recalled.

When the plane finally broke through the cloud cover, it was only 40 metres above the ocean. Alcock managed to gain control just in time to avoid plunging into the waves. Brown said they could hear the ocean "as its waves swelled, broke, and swelled again."

The temperature fell below freezing as the night wore on, causing ice to form on the plane's wings, fuselage, and controls. Icing increased the difficulty of steering the plane, and the additional weight meant that fuel consumption rose. The men, completely exposed to the elements in an open cockpit, also suffered from the cold.

With daybreak, Alcock brought the Vimy down to warmer air near the ocean surface in order to melt the ice. The engines stuttered from ice that remained lodged in the air intakes, but fortunately did not cut out altogether.

At 8:15 a.m. they caught sight of the coast of Ireland. Their ordeal was over and they knew they had achieved what no one had done before. After 16 hours and 27 minutes in the air, Alcock landed the plane on what appeared to be solid ground. It turned out to be a bog. The wheels of the aircraft sank and its nose plunged into the soft ground. Neither man suffered any injuries.

After being treated to a reception in Dublin, Alcock and Brown travelled to Britain by ship, arriving at Holyhead in Wales. Their heroics were rewarded with a triumphant procession through London, followed by the presentation of their £10,000 award by Winston Churchill, then Secretary of War. In addition to receiving the *Daily Mail* prize, Alcock and Brown were both knighted by King George V.

With Alcock and Brown's victory, Kerr abandoned his transatlantic quest. Instead, he and his crew ac-

British Cabinet Minister Winston Churchill (above left) presenting John Alcock and Arthur Whitten Brown with a cheque for the *Daily Mail* prize of £10,000. The airmen were greeted by large crowds in England after their aircraft crash-landed in Ireland (right).

After a harrowing overnight flight, lasting sixteen hours and twelve minutes, John Alcock guided the biplane onto what he thought was a grassy field at Clifden, Ireland (right). It turned out to be a bog, however, and the aircraft tipped nose-first into the ground.

cepted an invitation to fly to New York and tour the United States with the Handley Page machine. They flew to Parrsboro, Nova Scotia, on July 4, but a heavy landing on the makeshift airstrip destroyed the plane's fuselage.

The aircraft was repaired, and it reached New York in October. It was soon damaged again while making a forced landing at Cleveland, en route to Chicago from New York. At this point the tour was cancelled and Kerr returned to England.

Despite their triumph, both Alcock and Brown suffered misfortune in the months and years that followed. Alcock died six months later on December 18, 1919, when, while flying a plane from England to an air show in Paris, he crashed in northern France.

Brown only flew across the ocean one more time – to New York in 1946, to celebrate the commencement of commercial passenger flights between North America and Europe. His son Arthur, who flew with the Royal Air Force during the Second World War, had been killed during the D-Day invasion of France in 1944. Brown never recovered from this loss, and he committed suicide in 1948.

Just weeks after the success of Alcock and Brown, British airmen achieved another transatlantic milestone by making the first round-trip voyage in an airship. The *R-34* airship measured 196 metres, twice the length of a football field. Built during the waning months of the First World War, the *R-34* took off from East Fortune, Scotland, on July 2, 1919, on a round trip to New York.

A crew of 30 men and one stowaway, under the command of Royal Air Force officer Major George Herbert Scott, guided the hydrogen-filled airship across the ocean to Hazelhurst Field in Mineola, Long Island.

The most dangerous part of the journey came at the midway point of the North Atlantic, en route to Newfoundland, when the airship met stiff headwinds. As it neared the Newfoundland coast, the *R-34* was almost pushed back to sea by a series of violent storms. On July 4, Scott spotted some small islands lying off the east coast of Newfoundland. Brigadier General E.M. Maitland described the sighting of the islands as "the most thrilling moment of our voyage – successfully accomplished the first stage ... the first to bridge the gulf from east to west by way of the air."

The *R-34* had reached New-

foundland at Trinity Bay. A short time later, crossing Fortune Bay, the crew dropped a package of letters to a group of spectators. They reached Long Island on July 6, after four and a half days of flying. The massive airship delighted the large crowd that had gathered to watch it glide to a halt over Hazelhurst Field. Spectators were further thrilled by the spectacle of Major John Pritchard leaping from the passenger gondola with a mooring rope in hand and parachuting to the ground from an elevation of 600 metres.

After only three days in New Jersey, the *R-34* began its return flight to England, and reached Pulham in just over three days. In doing so, the *R-34* had made the first return crossing of the Atlantic Ocean.

Airships became popular with wealthy travellers during the 1920s and 1930s. Many lives were lost in accidents, however, and the perils of lighter-than-air flight would eventually force passenger airships from the skies. Among those who died in accidents were several *R-34* pioneers, including John Pritchard in 1921 and George Scott in 1930.

After the excitement of 1919, many people may have expected similar transatlantic heroics the following year. But those hopes would be dashed. If anything had been proven from the post-war competition between the US and Britain, it was that aircraft and navigation technology were too unreliable for long-distance flights over the treacherous North Atlantic.

Five years would pass before anyone attempted another ocean crossing.

On the austere coast of Labrador the sound of aircraft engines was heard for the first time in history on August 31, 1924, as a pair of American biplanes flew low over the icebergs and landed in the uninhabited harbour of Ice Tickle. The newspaper reporters watching from the deck of a naval ship, USS *Richmond*, were witnessing the final stages of the first successful flight around the world. Five months earlier, four seaplanes had taken off from Seattle, Washington, on this ambitious undertaking.

The US Navy organized and financed this round-the-world

The *R-34* airship completed the first return flight across the Atlantic Ocean in July 1919.

flight, just as it had in 1919 with the flying boats that crossed the Atlantic. The Douglas Aircraft Company was commissioned to design and build five planes for the venture. As well, American naval and diplomatic posts throughout the world had been ordered to support the flight, and spare parts had been delivered to several countries in advance of the planes' departure.

Four of the five World Cruisers built by Douglas prepared for the world flight, and were named for the American cities of Chicago, Seattle, New Orleans, and Boston. On March 31, the *Chicago*, *New Orleans*, *Seattle*, and *Boston* took off from a lake near Seattle, Washington. The contingent was soon reduced to three aircraft when the *Seattle* crashed in Alaska. The crew survived, but the aircraft was demolished.

Crew members of the Navy escort ship USS *Lawrence* (above) place a marker on Ice Tickle Island, Labrador, on August 31 where the aircraft first touched the Western Hemisphere after crossing the Atlantic Ocean. Spectators watch as the *New Orleans* is launched into Reykjavik harbour in Iceland (right) during the round-the-world voyage of the Douglas World Cruisers in 1924.

Governments throughout the world welcomed the three surviving World Cruisers, with the exception of the Soviet Union, which denied requests to allow them into its airspace. The planes were damaged numerous times, but logistical support from the US armed forces and embassies allowed them to keep going.

As the planes reached northern Europe it seemed as if all three would arrive triumphantly in the United States. But the North Atlantic would claim one of the aircraft. In mid-August, while it was flying between the Faroe and Orkney islands, the *Boston* encountered engine trouble and made an emergency landing at sea. A US Navy destroyer rescued the crew, but the *Boston* sank during a botched salvage attempt.

This left only the *Chicago* and the *New Orleans* to complete the world flight. These planes left Iceland on August 21 and arrived in Greenland the same day. Poor weather delayed their departure from the village of Ivigtut. On August 31, the weather cleared and the planes took off. They landed at Ice Tickle in Labrador following a 900-kilometre flight that took seven hours to complete.

Senior US Navy officers greeted Lieutenant Lowell Smith, commander of the *Chicago*, as he stepped ashore; "So that was it. We were in America," Smith recalled. "A launch came out from the *Richmond* and took us ashore after we had moved. We had landed after our travel, like Pilgrim Fathers on a large rock, but I'm afraid our first activities were not as edifying as theirs. We were just too darned happy for words."

The two aircraft departed Labrador on September 2, landing at Hawkes Bay on Newfoundland's Northern Peninsula. They arrived at Pictou, Nova Scotia, the next day, where they were joined by the only World Cruiser not taking part in the expedition – the prototype, named *Boston II*. In Pictou, the airmen were treated to a reception by the local chamber of commerce that included bagpipers and a lobster dinner. Later that day they were invited aboard the Canadian destroyer, *Patriot*, and fed lobster once again.

After leaving Nova Scotia, enthusiastic crowds then greeted the planes as they flew what amounted to a victory tour across the United States. On September 28 the 42,000-kilometre journey ended where it had begun six months earlier, near Seattle.

CHAPTER 2

While Charles Lindbergh made history by flying non-stop from New York to Paris in 1927, others that year were less fortunate. Mounting fatalities among transatlantic aviators convinced the crew of the *Royal Windsor* (seen here in Harbour Grace) to abandon their attempt.

ATLANTIC TRIUMPH, ATLANTIC DESPAIR

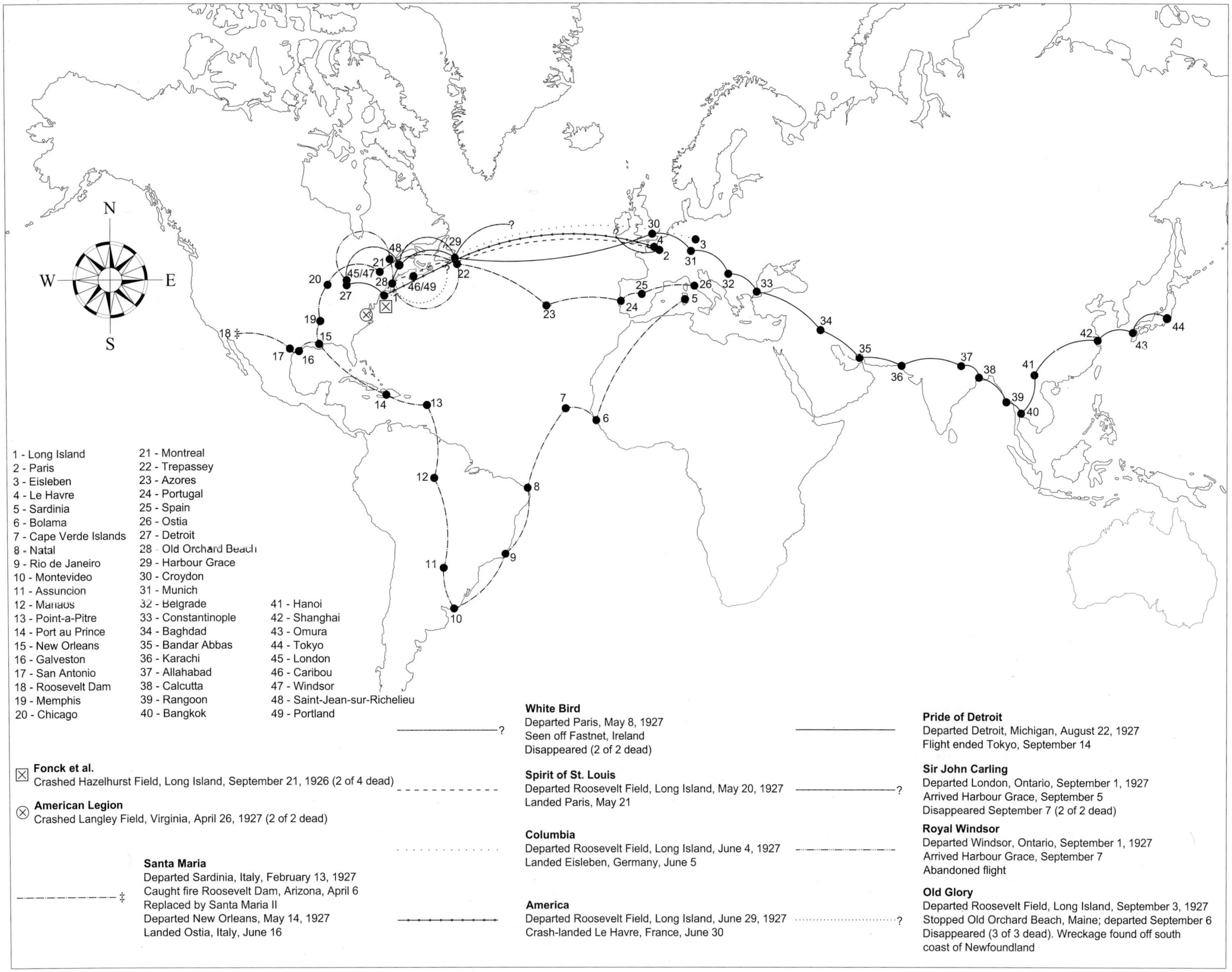

The glory of flying across the Atlantic Ocean tempted many aviators into the sky in 1927 – a year of triumph for a few, but one of disaster for many others.

Once again fame and prize money proved attractive lures, just as they had in 1919 when Alcock and Brown won the competition to cross the ocean by air. A few months after the Vickers Vimy crash-landed in Ireland, French hotelier Raymond Orteig offered US $25,000 to the winner of the first non-stop flight between Paris and New York.

At the time, aviation technology was too unreliable for such a lengthy journey, so nobody attempted to claim the prize until 1926, when French First World War flying ace René Fonck and three crewmen prepared to fly to Paris from Roosevelt Field, the renamed US Army Hazelhurst Field on Long Island. The term "field" was apt, as runways in this era were composed of clay and sand because paved surfaces were considered too hard for landing gear to absorb the shock of rough landings. Fonck and his crew planned to fly over Cape Breton, Nova Scotia, and Bonavista Bay, Newfoundland, before attempting to cross the treacherous waters of the North Atlantic. Tragically for them, part of the landing gear collapsed as their Sikorsky S-35 amphibious aircraft gained speed during takeoff. The plane, heavily laden with fuel, plunged over a bank at the end of the airstrip, and caught fire during takeoff. Fonck and navigator Lawrence Curtin survived, but two crewmen died.

Despite this fatal accident, several teams took part in the competition in 1927, starting with Noel Davis and Stanton Wooster, who intended to fly to Paris from Long Island in a tri-motor Keystone Pathfinder biplane named the *American Legion*. Davis, a former US Naval Air Corps aviator, and Wooster, who was still serving as a naval flyer, obtained permission from the US Army Air Services to use a Pathfinder for the transatlantic journey. The two men, who had financial backing from the American

Legion, tested the plane at an Army base in Virginia before embarking on their flight to Paris. The amount of fuel necessary to operate the plane's three motors had worried the pilots, however, since the machine weighed 7,700 kilograms when fully loaded. Despite this concern, on April 26 Davis and Wooster took the *American Legion* on its final test flight prior to attempting the transatlantic crossing. As the fuel-laden plane lumbered into the air, its pilots struggled with the controls to gain altitude. Instead of climbing, however, the aircraft lost speed as it cleared a grove of trees. Within seconds it careened earthward, diving nose-first into a shallow pond, instantly killing Davis and Wooster.

In France, Charles Nungesser and François Coli (far left) were celebrated in song, but the brave airmen disappeared over the North Atlantic. Ships and aircraft along the eastern seaboard, including a St. John's-based plane (above), searched in vain for the aviators. They were not the first to die in attempting a New York-Paris flight: two men perished when the *American Legion* (below) crashed, and two more were incinerated after their Sikorsky aircraft burst into flames during a test flight in Virginia.

This accident left three groups vying for the Orteig Prize, and a fourth – led by Richard Byrd – which was competing but did not officially register (and was therefore not eligible) for the prize.

By early May all four teams were testing their planes and training the men who hoped to make aviation history. On Long Island, three American planes were being outfitted. At Roosevelt Field Byrd's crew were frantically repairing their Fokker aircraft, the *America*, which had crashed during a test flight in April, injuring Byrd and pilot Floyd Bennett. At Curtiss Field, an airstrip adjacent Roosevelt Field, the *Columbia*, a plane owned by millionaire Charles Levine and designed by aircraft builder Giuseppe Bellanca, stood ready. Last to join the fray was Charles Lindbergh, a little-known barnstormer and airmail pilot, who hoped to fly solo to Paris in the *Spirit of St. Louis* and had rented a hanger adjacent Levine's at Curtiss Field.

Meanwhile, at Le Bourget Airport in Paris, public pressure was being exerted upon two French First World War aviators, pilot Charles Nungesser and navigator François Coli, to take to the air ahead of the American-led teams. Such fierce competition

The *Spirit of St. Louis* may have looked flimsy while being pushed onto Roosevelt Field (above), but it proved sturdy enough to carry Charles "Lucky" Lindbergh to Paris (above and right) in May 1927.

focused public attention and enthusiasm on the New York-Paris race, and newspapers on both continents were agog with speculation on who would win.

This frenzy of national pride induced some airmen to take unnecessary risks. In the race for the Orteig Prize, the French aviators allowed patriotism to take precedence over safety. The westward route planned by Nungesser and Coli, with Paris as their departure point, appealed to French public opinion, but the hazards were greater than an eastward crossing, due to the prevailing winds that flow west to east between North America and Europe.

After a rigorous series of test flights, Nungesser and Coli took off in their Levassseur PL8 biplane, the *Oiseau Blanc*, on May 8, despite forecasts of poor weather over the North Atlantic. Next day, French newspapers triumphantly reported that the *Oiseau Blanc* had landed in New York. Celebrations erupted throughout France.

But when confirmation failed to appear from North America, it became clear that this initial report was erroneous. As days passed without word of the fate of Nungesser and Coli, jubilation gave way to anxiety and, finally, despair. An extensive search by air over the ocean and along the coasts of Newfoundland and eastern North America turned up no evidence of the men or their plane.

This left the three US-led teams at Long Island to compete for the elusive prize. Many observers believed victory would go to Levine's single-engine mono-

Famed polar adventurer Richard Byrd lost his chance to win the New York-Paris race when his aircraft crashed during a test flight. Byrd (right) with pilot Floyd Bennett (at left) rebuilt the *America*. It crossed the Atlantic, but crash-landed on the coast of France.

plane, the *Columbia*, because, in 1926, it had set an endurance record by flying for 51 hours without refuelling.

The *Columbia*, which was prepared for the New York-Paris flight by Bellanca, a highly accomplished aviation engineer, would likely have taken off first if not for a rivalry between two pilots vying to fly the plane. The dispute, which was widely reported in newspapers, reached its climax when Levine selected Clarence Chamberlin; the losing pilot, Lloyd Bertaud, responded by obtaining a court injunction that grounded the plane.

This gave Lindbergh the time he needed to finish preparing his Ryan monoplane, *Spirit of St. Louis*. Lindbergh also benefited from the misfortune experienced by Byrd, whose three-engine aircraft required a new series of time-consuming test flights after it had been repaired.

On May 20, Lindbergh towed the *Spirit of St. Louis* to Roosevelt Field; he then climbed into its cockpit and taxied down the runway. The plane – heavily laden with fuel – strained to take off, lurched unsteadily into the air, and barely avoided telephone wires near the end of the runway.

Lindbergh flew northeast along the coast, crossing over Newfoundland's Avalon Peninsula before heading over the ocean. In his Pulitzer Prize-winning book, *The Spirit of St. Louis*, Lindbergh described the St. John's he saw as he passed over the city:

Charles Levine (left) was typical of the flamboyant risk-takers who financed and flew planes across the world's continents and oceans during the 1920s and 1930s. Levine is seen here with "Diamond Queen" Mabel Boll and pilot Bertrand Acosta.

I came upon it suddenly – the little city of St. John's, after skimming over the top of a creviced granite summit – flat-roofed houses and stores, nestled at the edge of a deep harbour.

It's almost completely surrounded by mountains. Further ahead, the entrance to the harbour is a narrow gap with sides running up steeply to the crest of a low coastal range, which holds back the ocean. Fishing boats are riding at buoys and men are at the wharves.

Twilight deepens as I plunge down into the valley. Mountains behind screen off the colours of the western sky. For me, this northern city is the last point on the last island of America – the end of land; the end of day ... North America and its islands are behind. Ireland is two thousand miles [3,200 kilometres] away.

The most dangerous part of the journey came between Newfoundland and Ireland, where Lindbergh flew through fog and darkness. Icing also plagued the flight, and at times Lindbergh flew the aircraft just a few metres above the ocean waves in order to keep ice from forming on the wings. Once in daylight again, weather conditions improved, and more than 100,000 people greeted Lindbergh when the *Spirit of St. Louis* landed in Paris. He had covered the 5,800-kilometre journey in 33.5 hours, thereby becoming the first person to fly solo across the Atlantic. Lindbergh became an instant idol throughout Europe and North America, and was given star treatment during an official tour of the US. His competitors at Roosevelt and Curtiss airfields now had to content themselves with lesser honours.

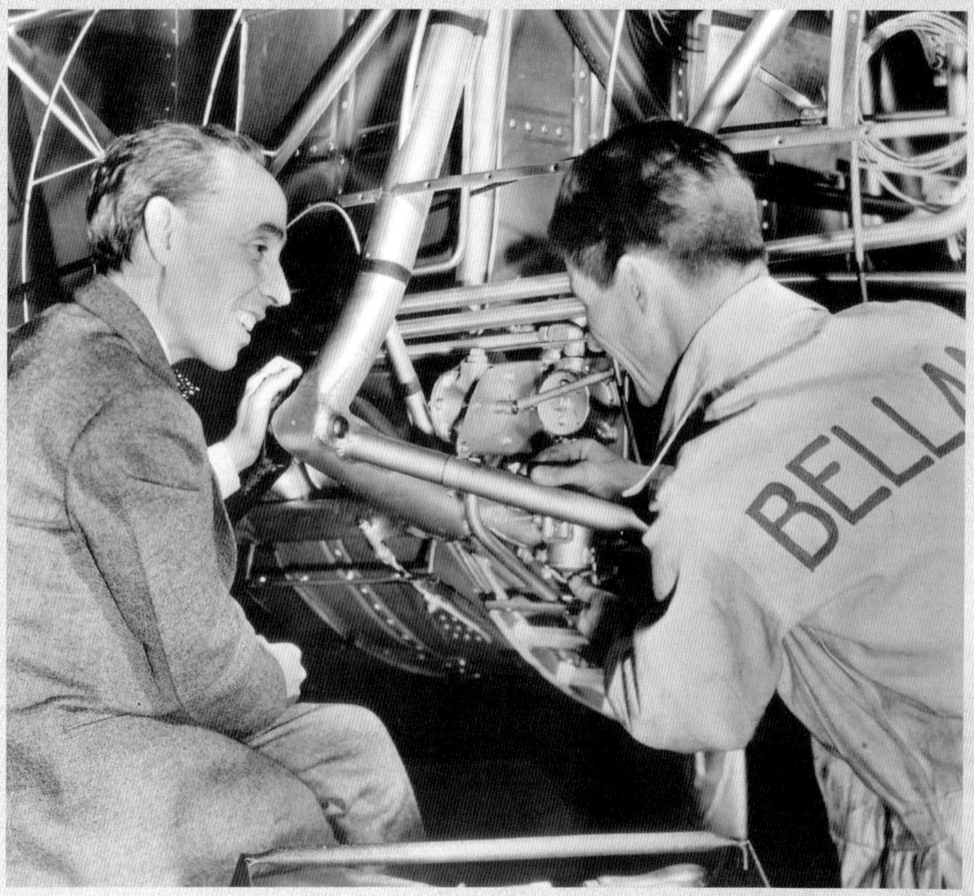

Giuseppe Bellanca (left, with a mechanic) was an Italian immigrant to the US who became one of the world's leading aircraft engineers. He designed the *Columbia* (seen below at Roosevelt Field in 1927), which lost the Orteig Prize to Charles Lindbergh. However, the *Columbia* was recognized as being a viable all-purpose aircraft rather than a plane built solely for a contest. Erroll Boyd, who in 1930 became the first Canadian to fly across the Atlantic, used the *Columbia* – renamed *Maple Leaf* – to reach Europe.

Just two weeks later, Chamberlin proved that the *Columbia* could have won the Orteig Prize if Levine had not antagonized Bertaud. The court injunction was lifted on the same day as Lindbergh's history-making flight, and on June 4 Levine, with Chamberlin at the controls, took off from New York, bound for Berlin. The aircraft followed Lindbergh's route until reaching the European coast, but it ran out of fuel short of the German capital. Levine's dream of landing in triumph among thousands of cheering fans evaporated when Chamberlin was forced to land the plane in the German town of Eisleban. By then, however, the *Columbia* had already broken the non-stop distance record by flying 6,300 kilometres, 500 kilometres more than Lindbergh.

The *America* finally took off for Paris on June 29 with Bernt Balchen and Bertrand Acosta as pilots, George Noville as engineer, and Byrd as navigator. Although he had been beaten by Lindbergh, Byrd used his flight for scientific purposes, to test a theory that changes in wind speed and direction occur at different altitudes. He wanted to determine whether aircraft could fly faster by taking advantage of such wind patterns.

Byrd proved the validity of this supposition during the *America* flight, as the aircraft made the fastest crossing yet of the Atlantic. Weather, however, would prevent Byrd and his crew from reaching Le Bourget Airport in Paris. Dense fog sent them off course over France, and the crew was forced to ditch the *America* into the ocean as it ran low on fuel. They landed without injury on the shoreline near Le Havre.

While much of North America and Europe revelled in the drama surrounding the New York-Paris race, Italians were engrossed by the feats of a pilot whose achievements had become a propaganda tool for Benito Mussolini's fascist

Francesco de Pinedo fought for Italy during the First World War. In 1927, dictator Benito Mussolini supported Pinedo's four-continent tour in a Savoia-Marchetti SM-55 flying boat. It was dubbed *Santa Maria*, the name given to the ship Christopher Columbus sailed to reach North America. The achievements of Pinedo and his crewmen, Carlo Del Prete and Vitale Zacchetti, were largely overshadowed by the publicity accorded Lindbergh.

The decade-long preeminence of Harbour Grace (above) as a port of call for transatlantic aviators ostensibly began with a world tour attempt by the *Pride of Detroit.*

regime. Francesco de Pinedo, an Italian air force pilot, became a national hero in 1925 after flying 54,700 kilometres over a period of several months in the South Pacific. This success would be followed in 1927 by a long-distance endurance test that offered even greater opportunities for demonstrating Italian flying prowess.

Pinedo and two crewmen, copilot Carlo Del Prete and mechanic Vitale Zacchetti, left Italy in a Savoia-Marchetti SM-55 flying boat, named *Santa Maria*, on a journey in 1927 that would cover over 46,600 kilometres and include four continents.

After leaving Italy, the *Santa Maria* touched down in Morocco, flew across the South Atlantic Ocean, and reached Brazil. The airmen travelled extensively throughout South America, including uncharted areas of the Amazon jungle, before turning north and flying to the US. When the *Santa Maria* arrived in New Orleans on March 29, it became the first foreign aircraft to fly into the United States.

The aviators then embarked on an extensive publicity tour of the US, with large crowds greeting the *Santa Maria* wherever it landed; but the marathon came to an abrupt halt when the aircraft caught fire during a refuelling stop at the Roosevelt Dam in Arizona. Nobody was injured in the blaze, which was the result of a carelessly discarded cigarette, but the famous plane was destroyed. This disaster did not put an end to Pinedo's journey, however. With Italy's national pride in the balance, Mussolini's Air Minister, General Italo Balbo, ordered the construction of an identical replacement plane.

Thus far, the Italians had been welcomed as heroes at every stop. But many Italian Americans opposed Mussolini's brutal regime and viewed Pinedo as a willing tool of the dictatorship. Simmering resentment climaxed into street violence as Pinedo supporters and anti-fascist groups fought during a public reception for the aviators on April 26, 1927, in New York.

When the replica, *Santa Maria II*, arrived in New York on board a ship in early May, Pinedo resumed his tour of the United States. By this time, however, the public had become absorbed by the New York-Paris race, which had already cost the lives of six men.

Because public attention was switching to competition for the Orteig Prize and away from the Italian marathon flight, Pinedo hoped to complete his return to Italy before a victor was declared in the New York-Paris competition. He intended to fly from Newfoundland to the Azores, and from there to Italy. Unfortunately for Pinedo, the *Santa Maria II* was unable to compete for the Orteig Prize because it lacked sufficient range to make a non-stop flight from North America to Europe.

Pinedo's bid to finish the tour before the Orteig Prize could be claimed was dashed, however, when poor weather grounded the *Santa Maria II*, first in Quebec and then in New Brunswick. On May 20, the day Pinedo flew into

The crew of the *Pride of Detroit* (from left): William Brock and Edward Schlee, with two unidentified men.

Trepassey harbour, Lindbergh took off from New York and was himself flying over Newfoundland on the way to his historic transatlantic flight. Pinedo did not see the *Spirit of St. Louis*, and left only a short, stark description of the barren landscape of the southern Avalon Peninsula surrounding Trepassey: "There was no sign of life but for a small fishing village."

The *Spirit of St. Louis* passed over St. John's while Pinedo was being driven to St. John's in the company of C. McKay Harvey, the Italian consul to Newfoundland. While in St. John's, Pinedo met numerous officials, and the government recognized his achievements by issuing overprinted postage stamps in his honour.

Pinedo soon returned to Trepassey, but rough waters delayed the departure of the *Santa Maria II* until May 23. The crew expected to fly directly to the Azores, but strong headwinds drained the aircraft of fuel and Pinedo was forced to land the seaplane in the vicinity of a fishing boat. The vessel towed the aircraft 320 kilometres to the Azores and, after the plane was repaired, Pinedo flew it back to the area where it had landed in the ocean. The crew then turned the *Santa Maria II* around and headed to Europe. Pinedo, Del Prete, and Zacchetti were greeted by cheering throngs of Italians when they landed in Ostia, west of Rome, on June 16, ending one of the longest flights to date.

Aircraft accidents later claimed the lives of two of the men who crewed the *Santa Maria II*: Del Prete, who died in a crash in 1928, and Pinedo, who was incinerated in 1933 when his plane caught fire during a takeoff from New York at the start of a planned non-stop flight to Baghdad.

Among the many people who were inspired by the Orteig Prize was Edward Schlee, the millionaire owner of Wayco Oil Company. Schlee was largely responsible for Harbour Grace becoming a focal point for early transatlantic aviators. An avid flyer, he had recently won an air race across the United States in a Stinson Detroiter monoplane.

Emboldened by this victory, Schlee hired pilot William Brock to join him in an attempt at setting a world record for circumnavigating the globe by air. They set an ambitious target of fifteen days to complete the journey, and Schlee decided that a stop in Newfoundland would be prudent in order to reduce the number of hours spent in the air between North America and Europe. He sent Fred Koehler, a representative of the Detroit-based Stinson Aircraft Corporation, to Newfoundland with a mission to find a suitable takeoff point and build an airstrip on it. While in Newfoundland, Koehler met with a group of Harbour Grace residents who convinced him to locate the airstrip in their Conception Bay town by offering public funding for a portion of the construction costs.

Thanks to financial assistance from the Newfoundland government and a donation from Stinson, the non-profit Harbour Grace Airport Trust Company built a gravel airstrip measuring 1,220 metres long and 61 metres wide. Over the next decade, this airfield would become the focus for transatlantic aviation until supplanted by Gander airport in the late 1930s.

With the airstrip completed, on August 25 Schlee and Brock flew their Stinson Detroiter, the *Pride of Detroit*, to Harbour Grace from Old Orchard Beach, Maine. One day later, the aircraft took off for Europe and landed safely at Croydon, England, after a terrifying flight. Nearing Ireland, the *Pride of Detroit* had encountered strong winds and rain that tossed the small plane about the night sky.

"The last few hours before dawn were our worst," Schlee wrote. "It seemed impossible that the airplane could live through such a storm. I kept thinking about my wife and two children in Detroit, wondering what they were doing. I even figured out what they probably were at that moment and I came to the conclusion they were doing what I was doing – praying."

Resuming their journey on September 1, Schlee and Brock reached Tokyo, Japan, two weeks later. Although their self-imposed fifteen-day deadline had passed, they decided to proceed with the last leg of their journey – crossing the formidable expanse of the Pacific Ocean. Traversing the Pacific by air was exceptionally hazardous in this era because navigational aids were relatively primitive and an aviator who became lost over the ocean would likely never be found alive.

Well aware of the risk, Schlee asked the US Navy to install a radio beacon on Midway Island, which would help him navigate from Tokyo to Honolulu. Meanwhile, American newspapers reported Schlee's complaint that the Navy had refused to provide him and Brock with charts. This public rebuke drew the ire of Navy

The *Sir John Carling* (below), with pilots James Medcalf (upper left) and Terence Tully, disappeared after departing from Harbour Grace.

Secretary Curtis Wilbur, who reminded Schlee that charts were available to anybody at a cost of 50 cents each. More critical to Schlee and Brock was Wilbur's refusal to use government funds for installing a navigational aid on Midway Island.

"As long as I have anything to do with the Navy, it will do nothing to aid and abet men to commit suicide," Wilbur told the *New York Times*. With this, Schlee and Brock loaded the *Pride of Detroit* onto a freighter and returned by ship to the US.

In the weeks following the exploits of Lindbergh, Byrd, Chamberlin, and the *Pride of Detroit*, several people, with varying degrees of experience, began plotting their own flights across the Atlantic. Despite the history of fatalities and accidents associated with transatlantic flights, many

The demise of the *Sir John Carling* and *Old Glory* convinced the crew of the *Royal Windsor* to stop at Harbour Grace and to abandon a transatlantic attempt.

Public euphoria over the transatlantic feats of Lindbergh, the *Columbia*, and the *America* soured after five aviators lost their lives within 24 hours in September 1927. *Old Glory* crashed into the ocean, killing all on board.

of them discounted the dangers posed by bad weather, unreliable equipment, and inadequate pilot training.

August and September would prove to be the deadliest months for transatlantic hopefuls. Eight men and women were killed in this period alone, intensifying the public backlash in North America against transoceanic flying. September began with two aircraft leaving southwestern Ontario on a marathon race to England. One of them, the *Sir John Carling*, was sponsored by the Carling O'Keefe brewery, which offered CAD $25,000 for the first Canadian crew to fly non-stop from London, Ontario, to London, England. The brewery had advertised widely for competitors, but was able to attract only one qualifier because of its stipulation that an entrant had to own his own plane. That aspirant, Philip Wood, was subsequently disqualified when organizers learned he was an American. Carling O'Keefe then opted to purchase its own Stinson Detroiter monoplane and hire pilots for the flight. Two Ontario Provincial Air Service pilots, Terence Tully and James Medcalf, were selected.

Rather than return home, Wood decided to compete against Carling O'Keefe. He found a sponsor in Windsor, Ontario, and convinced an Iowa-born resident of Toronto, Clarence Schiller, to be his co-pilot. They planned to fly non-stop from Windsor, Ontario, to Windsor, England. Like Tully and Medcalf, Schiller also flew with the Ontario Provincial Air Service.

Wood named his Stinson Detroiter monoplane *Royal Windsor*. On September 1, 1927, both the *Royal Windsor* and the *Sir John Carling* took off on their race to England. The idea of a non-stop flight quickly proved to be overly ambitious. Poor weather plagued the flights and both planes were forced to land far short of the British Isles. The *Royal Windsor* landed first near Montreal, Quebec, and then, two days later, at Portland, Maine. Thick fog forced Tully and Medcalf to make an unplanned landing of the *Sir John Carling* near Caribou, Maine.

With these delays, the sponsors of both flights were compelled to revise their non-stop rules. Tully and Medcalf decided to launch their transatlantic flight from Harbour Grace. Wood and Schiller opted to save time by skipping Newfoundland and flying non-stop to England from Maine. On September 5, after being delayed by fog in Maine, the *Sir John Carling* headed for the Newfoundland airstrip.

Wood and Schiller revised their plans once again, this time deciding to fly eight kilometres to Old Orchard Beach, where they hoped to cross the ocean with another transatlantic aircraft. This would-be companion was a single-engine Fokker monoplane named *Old Glory*. The flight of the *Old Glory* was the dream of newspaper magnate William Randolph Hearst. His plan was to finance the first non-stop flight from New York to Rome, in hopes of attracting more subscribers to his New York newspaper, the *Daily Mirror*.

The pilot chosen for this venture, Lloyd Bertaud, was best known for preventing the *Columbia* from competing for the Orteig Prize by obtaining a court order that forced Levine to keep it grounded. Hearst selected James Hill as navigator and co-pilot for the transatlantic flight to Rome. Hill, who was one of the most experienced aviators in the United States, had been working

THE NEW YORK TIMES, WEDNESDAY, SEPTEMBER 7, 1927.

TRANSATLANTIC PLANE OLD GLORY AND HER CREW.

International News Reel.

Old Glory on the beach at Old Orchard, Me., where she took off yesterday for Rome.

James Hill (left), passenger Philip Payne, and pilot Lloyd Bertaud.

OLD GLORY PILOTS ARE VETERAN FLIERS

Bertaud Made His First Hop in Glider When 12 Years Old, Crashing After 1,000 Feet.

HILL WAS STAR OF AIR MAIL

Made Night Flights for 3 Years in All Weathers Between Cleveland and Hadley Field.

Lloyd W. Bertaud and J. De Witt Hill, air mail pilots who are flying the big single-motored Fokker monoplane Old Glory in an attempt to reach Rome in a non-stop flight, are by temperament, experience and technical skill an ideal team for their task. Hill, 42 years old, lean, hard, quiet and studious, is, like Clarence Chamberlin, an engineer as well as an exceptional pilot. Bertaud, younger, more impulsive, equally skillful and resourceful, is an able complement to the older man.

For the last two years the experience of these pilots has been the best of training for their present assignment. Night pilots in the air mail between Hadley Field, N. J., and Cleveland must face fog, storm, mountains, cold, sleet and the fundamental problems of air navigation. Both of them have known what it is to have ice form on their wings and force them down within a few minutes.

Have Overcome Many Fogs.

Both have been lost in fog and above the clouds, where no beacon could reach them and with limited gas, slow ships and erratic compasses have found their way to safe landings many times. They feel these experiences qualify them for the transatlantic attempt better than long practice on the sea as navigators.

Bertaud was born at Alameda, Cal., in September, 1896. He doesn't recall when he first decided that he wanted to fly, but in 1908 when the Wright brothers here and Bleriot and Farman in France were startling the world with their first flights, as a twelve-year-old boy he aroused the enthusiasm of his playmates by constructing a glider with which he jumped off a cliff at Alameda Beach and flew nearly a 1,000 feet before he crashed.

Six years later he entered the Christofferson flying school at Ingleside Beach, Cal., and in February, 1914, received his pilot's license. A few months later the young flier met Lincoln Beachy at San Francisco and while he was learning from this pioneer stunt pilot he decided he would follow aviation as his life work.

Not long after his first meeting with Beachy, Bertaud stood on the ground with Hillary Beachy, the air-

as an airmail pilot.

Joining Bertaud and Hill was a passenger, Philip Payne, managing editor of the *Daily Mirror*. Canadian-born Payne had studied chemistry at the University of Toronto, and, after graduating, he moved to the United States, where he embarked on a career as a reporter. Like many of the journalists hired and promoted by Hearst, Payne was known for his tough investigative style. He was also impetuous, and tended to rebuff authority – qualities that would prove disastrous as the *Old Glory* was readied for its Atlantic journey.

Although billed as a non-stop flight to Rome from New York, the *Old Glory* actually stopped at Old Orchard Beach after leaving Roosevelt Field on September 3. The change was made due to fears of the pilots and the plane's builder, Anthony Fokker, who argued that Roosevelt Field was too short to allow the plane to take off safely with a full load of fuel, in addition to Payne, who added an extra 90 kilograms to its weight.

This seaside vacation spot of Old Orchard Beach, near Portland, had become a favoured departure point for aviators because it featured a natural runway: four kilometres of hard-packed beach sand. Here the *Old Glory* remained for three days while its crew waited for ideal weather conditions.

By the time the *Old Glory* had reached Maine, Hearst was reconsidering the whole venture. In mid-August, another Hearst-sponsored plane, the *Golden Eagle*, was lost in a race from California to Hawaii. With public criticism mounting over such "stunt" flights, Hearst did not relish the backlash that would result from losing a second plane and the three men it carried. More ominous yet was the demise of the *San Rafael*, a sister aircraft of the *Old Glory*, which disappeared in late August on a transatlantic flight from England.

On September 5, Hearst telegraphed Payne in Old Orchard Beach, imploring his editor and the *Old Glory*'s crew to turn back. He even pledged to pay Bertaud and Hill their entire fee if they cancelled the flight. But Hearst neglected to issue a direct order prohibiting Payne and the pilots from flying. In their enthusiasm, the three men were unmoved by intimations of disaster.

Shortly after midday on September 6, the *Old Glory* took off from Old Orchard Beach, just four hours before the *Royal Windsor* arrived from nearby Portland. Wood and Schiller had missed any chance they might have had to fly with the Hearst plane.

On board the *Old Glory* the pilots carried a wreath that they intended to drop into the ocean as a tribute to French pilots Nungesser and Coli, who had disappeared earlier that year. On the wreath was written, "Nungesser and Coli: You showed us the way. We followed. Bertaud, Payne and Hill." These words would be prophetic.

The plane was spotted several times as it crossed Nova Scotia, the Gulf of St. Lawrence, and the Burin Peninsula in Newfoundland. Although it was flying low, in the range of 90 to 300 metres, the *Old Glory* encountered ideal weather over the Gulf of St. Lawrence, and the men reported by radio that they were making good time. But something then went terribly wrong. Shortly after 4:00 a.m. on September 7, the *Old Glory* sent out an SOS call. Six minutes later, a second call came from the plane: "Five hours from Newfoundland, bound east."

Because Bertaud and Hill were using dead reckoning to gauge direction, they were unable to give rescuers a precise location. Several ships conducted a search, but nothing was found. Worse still, no further messages were received from the stricken plane. Within an hour of the distress calls, the *Royal Windsor* took off for Harbour Grace. Wood and Schiller had decided to follow the *Sir John Carling*, rather than attempt a direct flight to England from Maine.

It is not known whether Tully and Medcalf in the *Sir John Carling* had been told of the SOS call from the *Old Glory* or of the imminent arrival in Harbour Grace of their competitor, the *Royal Windsor*. By 7:00 a.m. on September 7 the *Sir John Carling* had taken off for England. The plane, which did not carry a radio, was never seen again.

These two disasters, combined with the recent deaths of another fifteen aviators over the Atlantic and Pacific oceans, resulted in calls for governments to outlaw such flights. Newspapers began referring to the ocean region between Newfoundland and Ireland as "The Atlantic Death Lane."

T. Douglas Robinson, Acting Secretary of the US Navy, declared that overseas flights were a "nuisance bothering everybody," and that they served no useful purpose. This public pressure compelled Wood and Schiller to abandon their quest. Hearst offered a reward for the recovery of wreckage from the *Old Glory*. No bodies were found, but on September 12 several large pieces of wreckage were retrieved 800 kilometres east of Cape Race by the Newfoundland steamship *Kyle*.

Press baron William Randolph Hearst's dream for boosting circulation for his newspaper empire turned into a nightmare when the crew of *Old Glory* plunged to their deaths off the coast of Newfoundland. The plane's wreckage was recovered by a Newfoundland coastal steamer, the *Kyle* (above and right). The bodies of the three men were never found.

When the *Sir John Carling* went missing, Carling O'Keefe officials claimed they had urged Tully and Medcalf to abandon their transatlantic flight, and had even offered to pay them the full $25,000 prize for some future flying achievement. According to this account, the aviators replied that their honour was at stake and declined to accept this alternative. After the two men died, Carling O'Keefe gave the prize money to the families of the dead flyers.

Although the *Royal Windsor* never flew across the Atlantic, its pilots made an indelible contribu-

tion during their stay in Newfoundland. While driving back to Harbour Grace from an excursion to St. John's, they and a group of journalists stopped at a forest fire that was blocking the road in the community of Kelligrews. The visitors began helping local volunteers fight the flames, which had engulfed several farmhouses. Wood, Schiller, and a photographer entered a burning home to rescue a blind man who was trapped in an upstairs room. After tying wet cloths over their faces, the three rescuers brought the man to safety.

While flying back to Ontario from Newfoundland, Wood and Schiller dropped a wreath into the ocean in memory of those who had died while attempting to fly across the Atlantic.

Schiller joined the Royal Canadian Air Force when war broke out in 1939, and flew bombers to Europe from Newfoundland with the Atlantic Ferry Command until his death, in a plane crash, in 1943.

Several pioneers of transatlantic flight led controversial lives, among them Byrd and Levine. As one of the most accomplished aviators of the post-First World War years, Byrd combined a talent for scientific innovation with patriotic zeal and a yearning for personal glory. Byrd's early achievements, which included crossing both the Atlantic Ocean and the North Pole by air, were largely tainted by disappointment and suggestions of fraud or incompetence. However, his reputation would later be salvaged by his dedication to exploring Antarctica.

During the First World War, Byrd demonstrated his technical ingenuity by designing instruments to assist seaplanes in making nighttime landings. He later pioneered air navigation technology that would become the foundation for polar exploration. After the war, Byrd turned to polar exploration, and in 1926 challenged Norwegian Roald Amundsen for the title of making the first flight over the North Pole. Amundsen, who had earned a stellar reputation by beating Robert Falcon Scott in an overland race to the South Pole in 1911, planned to fly over the North Pole in an Italian-built airship named the *Norge*.

Both teams established camps on Spitsbergen, a Norwegian island north of the Arctic Circle. Byrd hoped to make a return flight between Spitsbergen and the North Pole, while Amundsen planned to fly over the Pole to Alaska. Byrd and pilot Floyd Bennett took to the air first. Upon their return they claimed to have flown over the Pole.

Although Byrd reaped public adulation in the United States for this achievement, serious doubts were subsequently raised about the veracity of his claim. Many experts believed that the honour should belong to Amundsen, who had flown the *Norge* to Alaska following Byrd's alleged success. Amundsen died in a plane crash near Spitsbergen in 1928 while searching for survivors of the *Italia*, an airship which had broken apart in an accident, killing many of the crew and stranding several survivors.

A year later in the *America*, Byrd followed up his race to the North Pole by taking part in the New York-Paris competition. A series of delays, including an accident that injured Byrd and badly damaged his plane, forced him to postpone his attempt and gave Lindbergh the time he needed to carry out his historic flight.

Not only was the *America* beaten in the New York-Paris race but, as navigator, Byrd was primarily responsible for the aircraft becoming lost in thick fog over France. The pilots were eventually forced to ditch their plane in the ocean.

Byrd then turned to Antarctic exploration, a passion that consumed him for the rest of his life. In 1929 he made the first flight over the South Pole with pilot Bernt Balchen of Norway. Balchen had flown with Byrd on the *America* flight two years earlier. His five expeditions to Antarctica resulted in the discovery of thousands of square kilometres of territory, all of which were claimed for the United States. Byrd's dedication to the scientific study of Antarctica included spending the winter of 1934 in the interior of the frozen continent, where he collected data that became critical to the understanding of the climatic conditions of its inhospitable environment.

Charles Levine, one of Byrd's main competitors in the New York-Paris race, was one of the most colourful promoters of aviation in the 1920s. His life was marked by a mixture of greed, ambition, and courage.

The son of a New York scrap-metal dealer, Levine made his fortune after obtaining a contract with the US War Department in 1917 for the buying and disposal of shell casings.

In the mid-1920s, he and aircraft designer Giuseppe Bellanca formed Columbia Aircraft Company, which began manufactur-

ing planes. The *Columbia*, which Levine and pilot Clarence Chamberlin flew to Germany in 1927, was hailed as the most technologically advanced aircraft of its day.

Levine appeared to thrive on publicity, and frequently attracted controversy. In one notable instance, at a Chicago casino in 1928, Levine, who was accompanied by "Diamond Queen" Mabel Boll, punched a newspaper editor who had written an unfavourable article about him. The *New York Times* reported: "When the dust cleared, Miss Boll, sparkling with diamonds as usual, took Levine's arm and led him away." He lost heavily in the stock market crash of 1929, but continued spending money on experimental aircraft.

Trouble rarely strayed from Levine's side, but until the mid-1930s he managed to avoid significant legal difficulties. This changed in 1937, when he was convicted on charges of smuggling tungsten to the United States from Canada. He served two years in jail.

During the Second World War Levine would demonstrate a degree of courage and selflessness that had not hitherto been shown. In 1942, he was again convicted of smuggling, but this time it was for bringing a German citizen into the United States from Mexico. Even though the German turned out to be a refugee from a concentration camp, Levine was sentenced to 150 days in jail. The Columbia Aircraft Company was sold shortly after the Second World War, and Levine died in 1991 at the age of 94.

Until the mid-1930s, airships and fixed-wing aircraft vied for preeminence. This competition was exemplified in 1926 by a race between Richard Byrd and Roald Amundsen for the honour of being the first to fly over the North Pole. Byrd, who flew the aircraft (left), was credited with victory, but serious doubts later emerged about his claim. Amundsen flew an airship, the *Norge* (above), to Alaska from Spitsbergen Island.

CHAPTER 3

Glamour mixed easily with the drama of aviation in the 1920s. Socialite Mabel Boll (centre left) became the centre of attention in Harbour Grace when she landed there in 1928 with the intention of becoming the first woman to fly across the Atlantic. However, Amelia Earhart denied her that honour, and Boll is now largely forgotten.

WOMEN TAKE FLIGHT

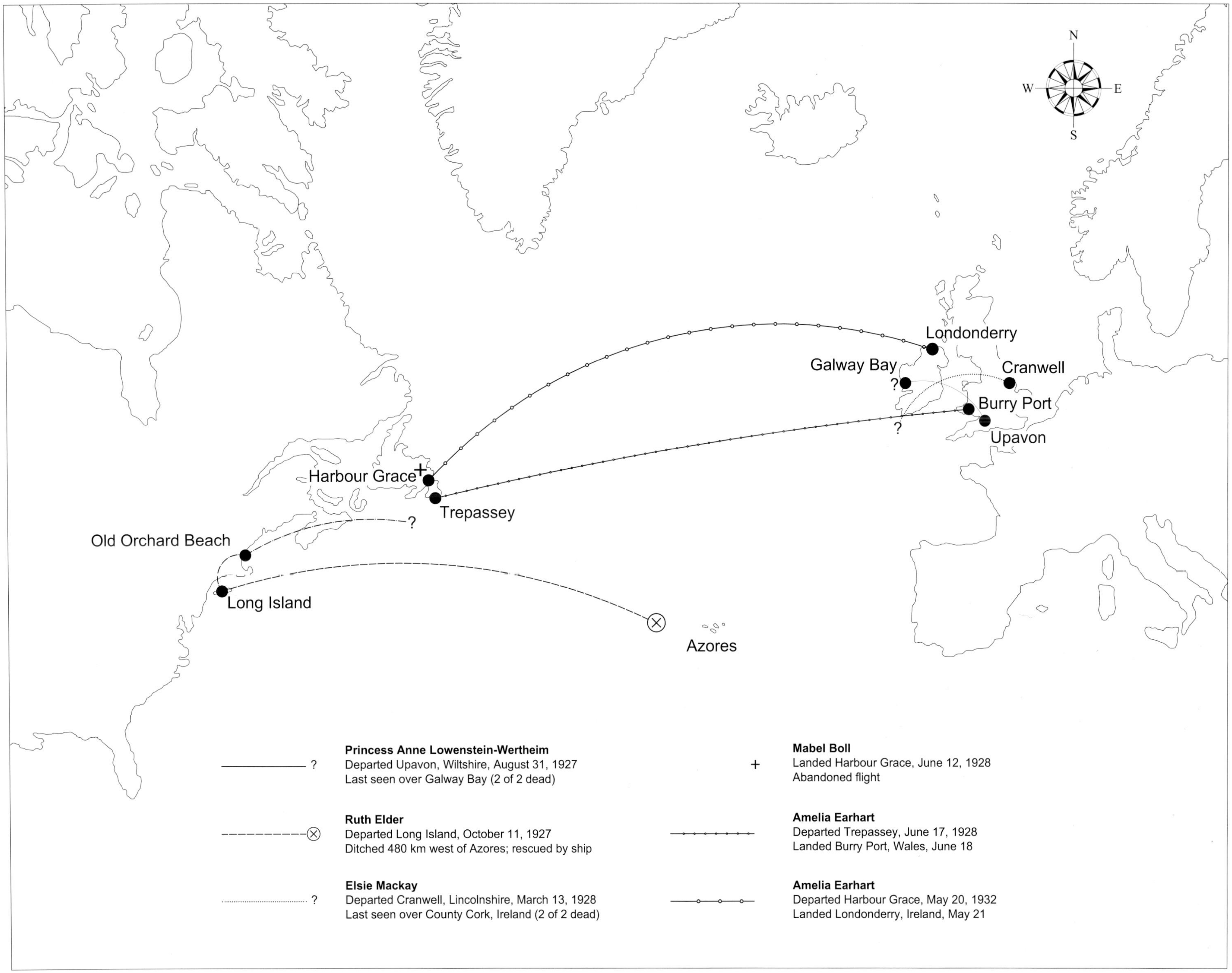

The swashbuckling image of aviators, personified by Charles Lindbergh, inspired many flyers with varying levels of experience to attempt their own transatlantic heroics in the months following his achievement. Many of these ventures ended in failure and, often, death.

Although men dominated the world of flight during the pioneer era, many women shared their dreams. In the mid-1920s and throughout the 1930s, they too would share in the glory and the tragedy. The goal of following Lindbergh and becoming the first female to make a non-stop crossing of the Atlantic lured several women to follow his lead. In August 1927, Anne Lowenstein-Wertheim, a wealthy, adventure-seeking 62-year-old aristocrat, made the first attempt. Born Anne Saville, daughter of the English Count of Mexborough, she became a princess by marrying a German prince, Ludwig Karl.

Her husband's early death allowed the princess the freedom to indulge her love for travel and aviation. In 1912, she became one of the first women to fly across the English Channel, and she later made long-distance flights throughout Europe and as far south as Egypt. In 1927, the princess decided to fly from England to Harbour Grace, and from there to Ottawa. She acquired a single-engine Fokker monoplane and hired two Imperial Airways pilots, Frederick Minchin and Leslie Hamilton, to be her crew. The aircraft, named *St. Raphael* (a patron saint of travellers), took off from Upavon, Wiltshire, on August 31 after being blessed by the Bishop of Cardiff. The crew of a ship spotted it flying slowly over Galway Bay, Ireland, and heading west. The *St. Raphael* and its crew were never seen again.

Princess Lowenstein-Wertheim (left) hired two pilots to fly her across the treacherous westward route across the Atlantic. Their aircraft was last spotted over the North Atlantic before it disappeared. Another victim of the North Atlantic was Frances Grayson (right), who died while flying a Sikorsky aircraft from New York to England. American pilot and film actress Ruth Elder (next page) was luckier: she and pilot George Haldeman were rescued from the ocean by a ship after ditching their plane during their transatlantic attempt.

In October 1927, a glamorous American film actress, Ruth Elder, made her attempt at flying across the Atlantic. Elder had limited flying experience, so she hired pilot George Haldeman to fly with her to Europe from New York. Elder's sponsors acquired a Stinson Detroiter aircraft, which she named *American Girl*. Unlike Lowenstein-Wertheim, Elder opted for the longer southern route, via the Azores, in order to avoid poor weather over the North Atlantic during the autumn months. However, a leaking fuel line forced Haldeman and Elder to ditch their aircraft into the ocean 480 kilometres west of the Azores. They landed in a shipping lane and were rescued. Despite coming short of their goal, Elder and Haldeman were awarded a tickertape parade in New York.

Elder's failure meant that there was now an opportunity for New York real estate agent and former newspaper reporter Frances Grayson, who had been preparing for several months to make the ocean crossing. Although Grayson had no experience as a pilot, Lindbergh's success had inspired her to obtain an aircraft and hire a crew to fly her to England from New York.

A strong believer in equal rights for women, Grayson was quoted as saying, "I am going to be the first woman across the

WRIGHT

Elsie Mackay (left) won plaudits for her acting, and notoriety for her lifestyle. After she disappeared during a transatlantic attempt in 1927, her co-pilot's widow, Emilie Hinchliffe (right), claimed she had received a message – through a medium – from her husband. Her daughter, Jean, is clutching a toy given to her by another aviator, Charles Levine.

bound for Harbour Grace. From there Grayson intended to complete what she expected would be a historic flight to London.

However, after leaving New York, and flying across the eastern seaboard of the US and Nova Scotia, the plane and its four occupants were never seen again. Reports suggested that the aircraft's engines were heard over the Newfoundland communities of Old Perlican, Harbour Breton, and Heart's Content, but these claims were never proved.

The mounting death toll among transatlantic aviators failed to deter a second English aristocrat from trying her luck at becoming the first woman to fly across the ocean. Not only did Elsie Mackay, daughter of the Earl of Inchcape, decide to fly the treacherous westward route, she chose to make her attempt in late winter, when weather conditions over the North Atlantic are particularly hazardous.

Mackay was a stage and film actress whose name had become tinged by scandal. In 1917, at the age of 24, she married a serviceman she had met while working as a hospital nurse. Lord Inchcape had objected to the marriage, and subsequently cut off all financial support to his daughter. Four years later the marriage was annulled after a highly publicized court case.

As Mackay learned to fly, she became known for her daring. While training for her pilot's

Atlantic, and mine the only ship since Lindbergh's to reach its destination. I will prove that woman can compete with man in his own undertakings."

With financial assistance from Aage Ancker, daughter of a Pittsburgh steel manufacturer, Grayson purchased a new Sikorsky amphibious aircraft, named *Dawn*. Several experienced pilots either declined to fly the plane, or left Grayson's team, particularly when the *Dawn* lost an engine during a test flight, and proved incapable of taking off with a full load of fuel. Eventually Oskar Omdal, a Norwegian, agreed to fly the plane, although Grayson planned to take turns at the controls. The crew also included navigator Brice Goldsborough, and radio engineer Frank Koehler, who had overseen preparation of the Harbour Grace airfield a few months earlier for the *Pride of Detroit*.

Rather than follow the safer but longer mid-Atlantic route taken by Elder, Grayson opted to gamble with the harsh winter conditions off the east coast of Newfoundland. After several months of delays the *Dawn* left Curtiss Field on December 23,

Mabel Boll, seen here in Harbour Grace, lost her bid to become the first woman to make a non-stop crossing by air to Europe from North America.

certificate in 1922, she had flown a plane so recklessly that her restraining belt snapped during an unplanned mid-air roll. If the instructor had not grabbed her, she would have fallen from the cockpit and plummeted to her death.

Lord Inchcape, who owned a large shipping firm, tried to rein in his thrill-seeking daughter by making her the artistic director of his steamship company. In this position she was responsible for cabin décor and kitchen design in her father's passenger vessels. She also became an advisor to the British Empire Air League.

But Mackay refused to settle down, and in the midst of international excitement over Lindbergh and his competitors she decided to become the first woman to fly across the Atlantic. In 1928, she hired an experienced pilot, Walter Hinchliffe, and gave him money to purchase a Stinson Detroiter aircraft. It was identical to the *Pride of Detroit*, which had flown to Europe from Newfoundland a year earlier. She named her plane *Endeavour* and planned to fly to Harbour Grace from England.

Mackay's father strongly opposed her intentions, so she publicly stated that Hinchliffe's friend, Gordon Sinclair, would accompany him. Mackay emphatically denied that she would fly in the *Endeavour*.

Only after the plane took off from Cranwell airport in Lincolnshire on March 13 did the public learn that Mackay, not Sinclair, was flying with Hinchliffe. The *Endeavour* was spotted flying over the coast of County Cork, Ireland, and then by ships at sea. It disappeared, however, and the wreckage was never found.

In November of that year, Hinchliffe's widow, Emilie, told an audience in London that, through a medium, she had received a message from her dead husband describing the last hours of his final flight.

"After leaving the Irish coast I flew in a northwesterly direction," she said he had told her. "We then got into the force of a storm. The gale was terrible. It broke one strut of the machine and another strut cracked. The canvass was tearing and one plug in the engine missing ... At three a.m. the machine came down within sight of the Azores. I trust some parts of the machine will be found near the rocks mentioned."

The notoriety that had dogged Elsie Mackay was to follow Lord Inchcape in the weeks following his daughter's death. To honour her memory he had established a

The *Friendship*, moored in Trepassey harbour.

More Butte News in the Standard Than in Any Other Paper

The Anaconda Standard.

MISS EARHART HOPS OFF OVER ATLANTIC

Oklahoma Tornado Kills Eight, Makes 300 Homeless, Causes Heavy Damage

ALL OF SOUTH MEXICO IS SHAKEN BY QUAKES PROPERTY DAMAGE BIG

MILLION LOSS ESTIMATED BY RELIEF CREWS

1,000 FLOOD SUFFERERS

Edwin Meredith Dies in Iowa

ON WAY ACROSS THE ATLANTIC

BOSTON GIRL AND TWO MALE MEMBERS OF CREW FAR FROM STARTING POINT AT MIDNIGHT

TOWNS ARE DESERTED

EDWIN MEREDITH FORMER CABINET SECRETARY DIES

HARRISON YOUTH MEETS DEATH IN PLANE ACCIDENT

MAN TAKES CHILDREN GOES TO WILDERNESS TO SPEND 40 NIGHTS

MONTANA STORM COMES IN TIME TO AID FARMERS

PRESIDENT COOLIDGE HEARS BLIND AND AGED PREACHER IN SMALL CHURCH AT BRULE

11 PERSONS MEET DEATH IN SUNDAY AUTO ACCIDENTS

£500,000 Elsie Mackay Trust Fund, which was to be administered by the British government. Emilie Hinchliffe then complained that Inchcape had refused her request for money to support herself and her three children. As public pressure mounted, Inchcape agreed to give her the equivalent of US $50,000.

This deadly series of failures by women to fly across the Atlantic would cnd in June 1928 with a showdown in Newfoundland between two very different personalities. One aspirant, Amelia Earhart, was already an accomplished, but little-known, pilot. Earhart was born in Kansas in 1897. She became entranced with aviation in 1917 while serving as a nursing assistant to servicemen recuperating at a hospital near an airfield in Toronto, Ontario. After earning her pilot's license in 1922 and purchasing a plane, Earhart set an altitude record of 4.2 kilometres and competed in several aircraft races. Financial difficulties forced her to sell the plane, although she found a job as a social worker and was able to support herself.

Amelia Earhart caused a sensation in Harbour Grace as she prepared her Lockheed Vega for a solo flight across the Atlantic in 1932. She and two reporters (above) discuss the upcoming flight. To her left are Lewis Esperance and mechanic Ed Gorski.

In 1927, while working in Boston, Earhart received an offer that gave her the chance to rejoin the world of aviation. An American heiress, Amy (Phipps) Guest, was funding what she hoped would be the first transatlantic flight by a woman. Guest, who had married a former British Air Minister, had obtained a tri-motor Fokker aircraft from Richard Byrd. She named it *Friendship*.

Guest hired publicist George Putnam to find two male crew who would fly the plane and a woman to travel with them as a passenger. She stipulated that the woman be American, a pilot, well-educated, physically attractive, and possess good manners. Earhart, who was known to Putnam, accepted the job, even though she would only be flying as a passenger.

On June 4, 1928, Earhart, along with pilot Wilmer Stultz – who had resigned as Grayson's pilot a year earlier – and mechanic Louis

In 1932 Amelia Earhart became the first woman, and the second person since Lindbergh, to fly solo across the Atlantic. She followed this achievement by setting speed and distance records in the Pacific Ocean. Earhart is seen atop her Lockheed Vega after completing a non-stop solo flight from Hawaii to Oakland, California, in 1935.

Officers of the passenger liner, *President Roosevelt*, with Louis Gordon (second from right) and Wilmer Stultz (fourth from right), listen as Earhart reads a congratulatory telegram during their triumphant return to the United States.

Gordon, landed in Trepassey, Newfoundland. They had started the journey in Old Orchard Beach, Maine, and stopped at Halifax, Nova Scotia, en route. The trio expected to take off for Europe from Trepassey the day after their arrival, but high winds and fog delayed their departure. They would become guests of the people of Trepassey for nearly two weeks.

Competing with Earhart was flamboyant New York socialite and cabaret performer, Mabel Boll. Known as "The Diamond Queen," Boll was romantically involved with millionaire playboy Charles Levine. Levine had already set an aviation record in 1927 when he flew as a passenger with pilot Clarence Chamberlin from New York to Germany in the *Columbia*. In March 1928, Boll and Levine had flown as passengers with the *Columbia* on the first non-stop flight to Havana, Cuba, from New York.

Levine agreed to loan Boll the *Columbia* for a transatlantic flight on which she would be a passenger. Boll had expected Stultz to be her pilot, since he had flown her and Levine on the record-setting Havana flight. She became incensed when Stultz agreed to fly with Earhart and the *Friendship* instead. "I can't understand it," Boll complained. "Now he has taken off with this other woman, and I was sure he would fly with me. I depended on him."

Boll responded by hiring a replacement pilot, Oliver Le Boutillier, and a mechanic, Arthur Argles. She had been delayed in starting her flight from New York, but, fortunately for her, poor weather in Trepassey had grounded Earhart. The *Columbia*, with Boll and her crew, flew to Harbour Grace from New York on June 12.

Like Earhart, Boll anticipated only a short stop in Harbour Grace before taking off for Europe. However, the fog and high winds that grounded the *Friendship* also prevented the *Columbia* from leaving. Boll was compelled to find other activities to keep her busy until the skies cleared. The Knights of Columbus threw a reception for the crew of the *Columbia*, and they were fêted by dignitaries during a tour of St. John's.

In contrast, Earhart, Stultz, and Gordon never strayed far from their aircraft. Having been unprepared for a lengthy stay in Trepassey, and arriving with only the clothes they were wearing, the three aviators were forced to buy or borrow clothing from local residents.

Earhart wrote that "[t]he flannel nightgown ... was borrowed and I began to feel that even its sturdy fabric would be worn out before we ever got away from Trepassey – although I didn't know about the wearing qualities of flannel gowns, never having had one before. Incidentally, its warmth was supplemented by the down beds upon which we slept and into which we sank luxuriously."

Their time in Trepassey was largely spent playing cards, fishing, and sightseeing. Music proved to be an activity that brought visitors and residents together, and Earhart described how local people would gather outside the fence of their boarding house whenever an instrument was being played.

Finally, after several aborted attempts at taking off, the *Friendship* rose from Trepassey harbour on June 17. Under the command of Stultz, and carrying only essential equipment and little food or water, the plane flew across the North Atlantic with Ireland as its destination.

Fog shrouded the *Friendship* throughout much of the trip, and navigation became more difficult after its radio broke down. The plane became lost as they neared Europe and, with only one hour of fuel remaining, they decided to land at the first opportunity. This happened to be Burry Port, a coastal town in Wales. The initial reaction to the *Friendship* from local residents was mild curiosity. This changed to frenzied excitement once word spread that the plane had crossed the Atlantic and that one of its occupants was a woman.

When news arrived that the *Friendship* had taken off, and that Earhart had won the race to be the first woman to cross the ocean by air, the disappointment was too much for Boll. She donated US $500 to the operators of the Harbour Grace airstrip and flew with the *Columbia* back to New York.

After the successful flight by the *Friendship*, the careers of the two women veered in radically different directions. Unlike Earhart, whose flight gave her an opportunity to proceed with a career in aviation, Boll soon lost interest in flying. For a few years, her tempestuous lifestyle and associations with wealthy men such as Levine kept her name in New York newspapers. Boll would marry five times throughout her life; her husbands included a South African plantation owner, a Polish count, and a musician with the New York Philharmonic Orchestra. Boll, who was rarely seen in public without her diamonds, once conceded that she loved jewels more than men. "There is no pleasure in the world comparable to the possession of diamonds," she said. Boll suffered tragedy in 1942 when her only child, a son, died of appendicitis. In 1949 she was committed to a Manhattan mental hospital. She died there that year, apparently of a stroke.

In contrast with Boll, Earhart's celebrity status continued to grow. In 1928, the public in both North America and Europe clamoured to learn more about the woman who had conquered the Atlantic. She was dubbed "Lady Lindy," in part due to her physical resemblance to Lindbergh.

But Earhart felt uncomfortable taking credit for the flight. She downplayed her role and emphasized that Stultz and Gordon had actually flown the plane. "I was a passenger on the journey – just a passenger. Everything that was done to bring us across was done by Wilmer Stultz and 'Slim' Gordon," Earhart wrote. "Any praise I can give them they ought to have."

Earhart would later note that while Stultz received payment of US $20,000 and Gordon US $5,000, she was required to turn over the fees she received from writing accounts of the journey to the flight's promoters.

This sense of having unfinished business spurred Earhart to prepare for another crossing of the Atlantic in 1932, but this time with herself at the controls. This flight, and other achievements throughout the mid-1930s, would establish her as an icon of human achievement.

Earhart was determined to demonstrate that female pilots were as capable as men at flying. To prove her point, she would mark the fifth anniversary of Lindbergh's 1927 New York-Paris flight by making her own solo transatlantic journey to Paris.

In the early hours of May 20 – the anniversary of Lindbergh's transatlantic flight – Bernt Balchen flew Earhart and mechanic Ed Gorski to Harbour Grace from Roosevelt Field, Long Island. Balchen, who had gained valuable experience flying across the Atlantic with Richard Byrd in 1927, assisted Earhart in preparing for her own solo journey.

Earhart slept five hours at Empire House, a Harbour Grace hotel, before climbing into the cockpit of her single-engine Lockheed Vega. Hotel proprietor, Rose Archibald, gave her a thermos of soup moments before she took off for Paris.

Difficulties plagued her throughout the journey, starting with failure of the altimeter. She then flew into a severe lightning storm and was pushed off course. More serious yet, a weld in the exhaust manifold split open. During the night, flames from fuel combustion shot brilliantly into the darkness.

Later, the Vega flew into dense cloud. As she climbed to escape, a film of ice appeared on the aircraft's wings. Then, the plane suddenly spiralled down to the ocean. As it reached the warmer air below, the ice melted, allowing Earhart to regain control and pull the plane out of its dive.

"How long I spun I do not know," she wrote. "As we righted and held level again, through the blackness I could see the whitecaps, too close for comfort."

More unnerving yet, fuel began leaking onto her shoulder from the cabin fuel gauge. This leak, combined with the widening split in the exhaust manifold, posed a significant fire risk. Earhart decided to land at the first opportunity rather than continue on to Paris. She landed in a pasture near Londonderry, Northern Ireland.

Earhart came short of reaching France, but to the public that did not matter. Huge crowds followed her in London after she flew to the British capital from Ireland. Unlike many pilots of her era, Earhart downplayed the importance of the flight to the progress of aviation. Instead, she used her achievement to champion women's rights.

"Outside of demonstrating that a woman can fly the Atlantic alone, I don't see that I have added anything to the science of aviation or anything else," she told journalists in London. "But if I've helped break through the barriers of the tradition that women are helpless, that's something, isn't it?"

Earhart's feminism was radical for her time, and she used her status to advance her philosophy. In September 1932 Earhart was part of a delegation from the National Woman's Party that met President Herbert Hoover to urge his support for legislation guaranteeing equal rights to women. She also believed that women are, in some ways, physically stronger than men – a theory tested, and partially proven, during the early years of the American space program. "Woman really is capable of standing strain better and longer than man," Earhart said.

Earhart looked forward to the day when she would fly on passenger aircraft across the world's oceans. She would never realize that dream. In 1937, she and navigator Fred Noonan took off from Oakland, California, on a round-the-world journey. They followed a route close to the equator, and hoped to better Wiley Post's time, which had been set in 1931.

Flying east, they arrived in New Guinea on June 29. From there Earhart planned to fly to tiny Howland Island 3,500 kilometres away, the longest overwater leg of the trip. She and Noonan were unable to find the island, and they disappeared. Numerous attempts have been made in the ensuing decades to find the wreckage of Earhart's plane; none has been successful.

Aviation royalty in Atlantic City, New Jersey, on August 2, 1933. A swimming party was given for husband and wife flyers, Amy Johnson (seated, right) and James Mollison (right), by Amelia Earhart and her husband, George Putnam (left). Mollison had recently suffered minor injuries after he and Johnson crash-landed their plane after achieving a successful westerly crossing of the Atlantic.

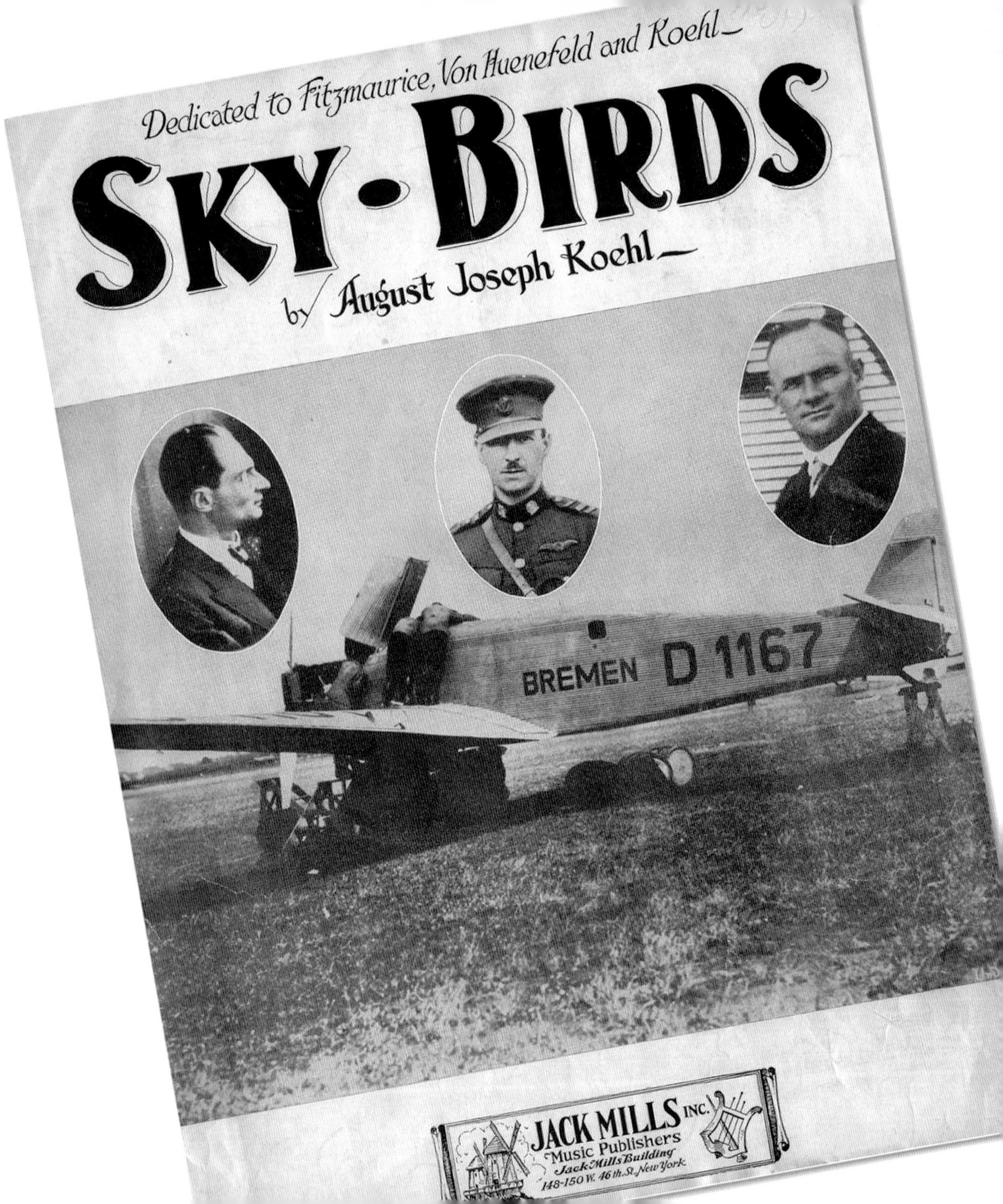

Composers sometimes wrote music to celebrate well-known aviators, in this case the crew of the *Bremen*, which, in 1928, achieved the first westerly crossing of the Atlantic.

THE LUCKY AND THE FOOLHARDY

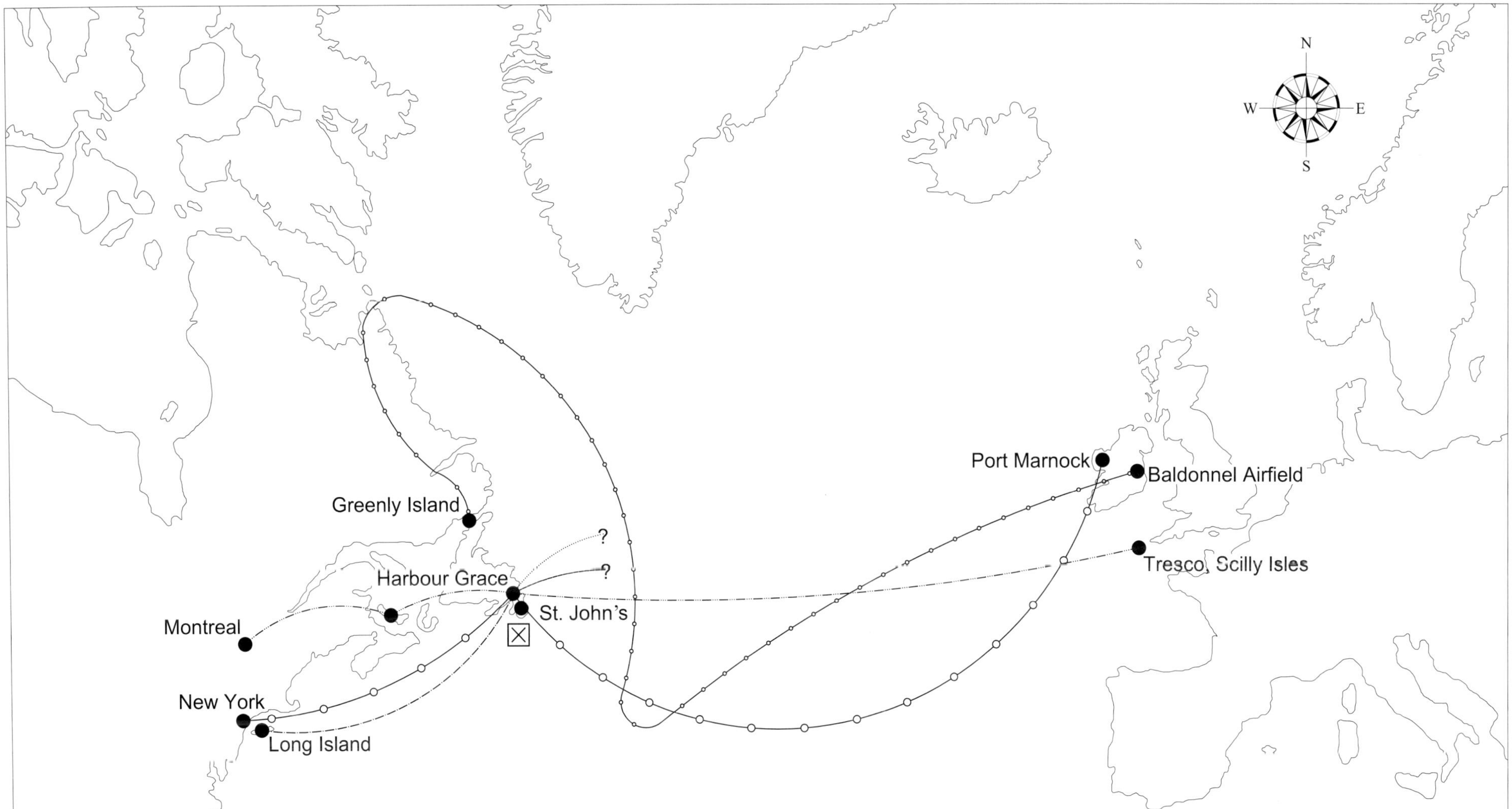

Bremen
Departed Baldonnel Airfield, Ireland, April 12, 1928
Landed Greenly Island, April 13

Harold Macdonald ?
Departed Harbour Grace, October 23, 1928
Disappeared (1 of 1 dead)

Urban Diteman ?
Departed Harbour Grace, October 22, 1929
Disappeared (1 of 1 dead)

Southern Cross
Departed Port Marnock, Ireland, June 24, 1930
Landed Harbour Grace
Arrived New York, June 26

Charles Wynne-Eyton
Crashed on takeoff from St. John's, July 6, 1930

City of New York
Departed Long Island, August 2, 1930
Landed Harbour Grace
Crashed on takeoff, August 3

Maple Leaf
Departed Montreal & landed Prince Edward Island, September 10, 1930
Landed Harbour Grace, September 23
Departed October 10 & landed Tresco, Scilly Isles

If anybody had believed that the carnage of 1927 would dampen enthusiasm among aviators for flying across the Atlantic, they were mistaken. Pilots in North America and Europe watched in envy as Charles Lindbergh basked in the glow of public adulation and grew rich in an avalanche of endorsement contracts.

Many aviators dismissed the catastrophes that followed Lindbergh as being the products of bad luck, inexperience, or poor planning. In the spring of 1928, a German-led expedition would attempt to fly the deadly east-west route across the Atlantic. Less than a year had passed since several flyers disappeared going this direction, among them Charles Nungesser and Elsie Mackay.

Aircraft flying from Europe to North America must contend with powerful prevailing winds that flow west to east across the Atlantic, conditions that slow progress through the skies and force planes to use more fuel. But these risks did not deter Baron Guenther von Hünefeld, a wealthy German aristocrat, who had been inspired by Lindbergh's New York-Paris flight.

Hünefeld now wanted to become the first person to fly westward across the ocean. He purchased a Junkers W33 and named it *Bremen*. Hünefeld was not a pilot, however, so he hired Hermann Köhl, a former First World War airman who had become chief pilot for Luft Hansa (which was renamed Lufthansa in 1933). This 1928 bid was not their first attempt at crossing the Atlantic. In August 1927, Köhl and Hünefeld battled strong winds for 24 hours before abandoning the flight and returning to Germany.

They waited until the following year to make another attempt. On March 26 the *Bremen* flew to Baldonnel Airfield, an airfield near Dublin. Here, Köhl and Hünefeld were joined by a senior officer in Ireland's air force, James Fitzmaurice.

Fitzmaurice had learned to fly during the First World War. He served with the Royal Air Force, but joined the Irish air corps in 1922 after Ireland gained independence from Great Britain. Like Köhl and Hünefeld, Fitzmaurice had tried to make a transatlantic flight in 1927. In September that year, he and co-pilot Robert McIntosh had taken off from Ireland in a Fokker monoplane named *Princess Xenia*. Strong winds forced them to turn back.

Köhl, Hünefeld, and Fitzmaurice may have wondered whether history was being repeated as they waited for skies to clear over the North Atlantic in March and April 1928. Poor weather kept the *Bremen* grounded for seventeen days at Baldonnel Airfield until the early morning of April 12 when conditions were judged favourable for flying across the ocean. The *Bremen* rose into the sky and set a course for Mitchel Field, an airfield in Garden City, Long Island.

The *Bremen* would never reach New York. Nothing was heard from the plane after it cleared the coast of Ireland. A radio had not been brought on board because the added weight would have increased fuel consumption. As the hours passed, it seemed as though the *Bremen* had disappeared like so many of its predecessors.

But then, after two days of silence from the missing plane, a telegraph message was transmitted to the world from Point Amour, a village located on the south coast of Labrador. On April 13, after 35 hours in the air, the plane had crash-landed at Greenly Island, Quebec, a barren, treeless outcrop in the Strait of Belle Isle close to Labrador and Newfoundland.

All three men aboard the plane emerged uninjured after the *Bremen* landed on a frozen pond

and crashed through the ice. Their story of flying blindly through fog and stormy weather – and of how they risked death in landing on a rocky island – would captivate the public through newspapers, magazines, radio broadcasts, and newsreels.

According to the men, the flight had gone well until nightfall, as the *Bremen* approached the Grand Banks area east of Newfoundland. Here, the plane encountered a massive fog bank. Not only were Köhl and Fitzmaurice unable to guide the aircraft by the stars, but in the fading light they found it difficult to use the compass.

The pilots fought against wind, rain, snow, and fog as they flew blindly through the night. As the sun rose, the fog began to clear and the three men found themselves travelling along the coast of northern Labrador, heading into the Arctic. Realizing that they were off course, they turned the *Bremen* south and followed the Torngat Mountains and Lake Melville.

At about noon, through a gap in the cloud cover, they sighted what appeared to be a ship. By now, the *Bremen* was running low on fuel. Despite the danger, they decided to land in the water and await rescue. Upon descending, however, the men discovered that what they were looking at was actually a lighthouse on Greenly Island.

In their desperate situation, Köhl and Fitzmaurice had no choice but to risk death by crash-landing. As the plane landed, the wheels sank into a peat bog, damaging the aircraft beyond repair: but the men were safe.

The flyers soon received excellent care from a group of seven families who were living on Greenly Island and working at a fishing station owned by Job Brothers, a Newfoundland fish trading company. Telegraph stations on the south coast of Labrador were alerted to the arrival of the *Bremen*. News of the flyers' landing quickly spread around the world, and a race to Greenly Island by the largest newspapers in North America immediately ensued. But getting to the island proved to be almost as difficult as the flight of the *Bremen* itself. Pack ice had blocked the Strait of Belle Isle, and even a Canadian icebreaker was unable to reach the island.

On April 15, a flight sponsored by the *Toronto Daily Star* was the first to arrive on Greenly Island. Its pilot, Clarence Schiller, had flown with Robert Wood on an aborted Atlantic crossing by the *Royal Windsor* in 1927. Schiller, who was acting

The crash-landing of the **Bremen** on Greenly Island (previous page) set a record, but also left the aircraft too badly damaged to take off again. The flyers – (left to right) Baron Guenther von Hünefeld, James Fitzmaurice, and Hermann Köhl – escaped without injury and were fêted in a tickertape parade in New York after their rescue.

Harold MacDonald (right), seen here with Harbour Grace airstrip supervisor Harold Archibald (centre), suffered the same fate as many transatlantic aviators during the 1920s and 1930s and was never seen again after taking off from Harbour Grace.

as a reporter for the *Star*, was among 60 journalists who would make the difficult journey to Greenly Island.

One of those caught in the media frenzy was Floyd Bennett, the pilot credited with flying Richard Byrd over the North Pole in 1926. Bennett and Bernt Balchen, who flew the *America* in a transatlantic crossing in 1927, were hired by the *New York World* newspaper to fly spare parts to repair the *Bremen*.

Bennett had been recuperating in hospital following a recent air crash when he was asked to fly to Greenly Island. Despite his fragile condition, he agreed. But during the 1,600-kilometre flight he contracted pneumonia and was rushed to a hospital in Quebec City. As Bennett's condition worsened, Charles Lindbergh flew non-stop to Quebec City from New York, carrying medicine to save his life. But Lindbergh was too late to save his friend. As a tribute, Byrd named the plane that took him over the South Pole in 1929 after Bennett. An airport in Brooklyn, New York, was also named for Bennett.

At Greenly Island, attempts at repairing the *Bremen* were unsuccessful. Köhl, Hünefeld, and Fitzmaurice were finally flown from the island, almost two weeks after they had landed. A tickertape parade held for the flyers four days later attracted two million people along a 16-kilometre route. They followed

Harold MacDonald preparing for his transatlanic flight.

this with a tour of the American northeast, as well as Montreal, Quebec City, Germany, and Ireland. Hünefeld, who completed a flight throughout Southeast Asia in September later that year, died in 1929 from tuberculosis.

The fate of the *Bremen* differed markedly from most transatlantic bids that had gone wrong in that the flyers were rescued. Others were not so lucky, including many who had followed the *Pride of Detroit* by using Harbour Grace as their departure point for crossing the ocean by air. However, the dangers of attempting this feat are illustrated by a sobering statistic: by mid-1928 only one of the five aircraft that had used Harbour Grace as a departure point for Europe had been successful. The *Pride of Detroit* managed to reach England, but the *Sir John Carling* disappeared. Three other flights – Admiral Mark Kerr's *Atlantic* in 1919, the *Royal Windsor* in 1927, and Mabel Boll's *Columbia* in June 1928 – had been abandoned. However, in October 1928 Harbour Grace would host another bid for a flight across the Atlantic.

Harold MacDonald, an Englishman who had served in the Royal Navy during the First World War, arrived by ship in St. John's on September 23. MacDonald, who had recently retired from the navy, had only learned to fly a year earlier. Immediately after obtaining his pilot's licence MacDonald had embarked upon a flight from Great Britain to India, despite accumulating only ten hours' experience flying solo.

The European leg of the journey was uneventful, but MacDonald's luck changed once he entered the Middle East. His plane was damaged in an accident while landing in Baghdad. Then, after repairs were completed, he crashed once again in the Sahara Desert. An Arab group took MacDonald into custody, but released him following diplomatic intervention by the Italian military. MacDonald abandoned his planned journey to India and returned home.

Despite the accidents and limited flying experience, MacDonald began planning for a transatlantic crossing. The success of Lindbergh, combined with patriotism, inspired him to make the perilous journey. MacDonald decided to use a small de Havilland Gypsy Moth biplane, which was the same open-cockpit aircraft he had used the previous year on his aborted voyage to India. The single-engine craft generated 85 horsepower, just a fraction of the 220 horsepower produced by Lindbergh's *Spirit of St. Louis*. By crossing the Atlantic in the Gypsy Moth, MacDonald planned to set a record for flying the smallest aircraft across the ocean.

Another solo flyer, American Urban Diteman, greatly underestimated the perils of flying across the Atlantic Ocean. He and the *Golden Hind*, which was named after one of the ships commanded by sixteenth-century English adventurer, Sir Francis Drake, disappeared.

Poor weather conditions forced one of Australia's greatest aviators to pay an unexpected call on Harbour Grace (above and right) during a round-the-world voyage by the *Southern Cross* in 1930. Charles Kingsford-Smith (above centre) flew his Fokker F-VIIb on several record-breaking flights, until his death in 1935.

MacDonald did not appear to appreciate the dangers of flying across the Atlantic. The plane's open cockpit left him unprotected from the cold autumn weather; but, more critically, he had only a half-hour experience of night flying and had never flown over a huge expanse of ocean with no landmarks to guide him.

In a parting letter to his wife, MacDonald justified his decision to make the journey: "I know you are against the flight. I know that people will say that I am a suicidal fool, but I do not think so. After all, they called Lindbergh 'the flying fool'. If an American can do it, then so can an Englishman."

In the letter MacDonald also acknowledged that he would possibly never see her or their five-year-old son again, and furthermore, if he died, she should allow the boy to join the army if he were to show an interest. He then apologized to her: "If I have made a mistake, I am sorry. It will be terrible for you."

His plane was sent to Harbour Grace from St. John's by train, where it was reassembled for the flight to England. MacDonald took off on October 23 after waiting three weeks for suitable weather. Before leaving, MacDonald confidently predicted that he would meet his wife the next day. He also listed his provisions for the flight: "I am taking with me a quart thermos bottle of coffee and two dozen sandwiches. I hope to eat the balance of the sandwiches in London tomorrow."

He had difficulty lifting the plane into the air from the airfield because of the heavy load of gasoline on board. A Dutch ship, *Hardenburg*, spotted MacDonald's plane about 960 kilometres east of Cape Race, Newfoundland. The Gypsy Moth then disappeared. No wreckage was found.

MacDonald's solo effort would not be the last flight to end in tragedy. In October 1929 the delusions of a Montana cattle rancher led to his lonely demise in the Atlantic Ocean. Urban F. Diteman believed that he was a direct descendant of Sir Francis Drake, the sixteenth-century English privateer and vice-admiral who circumnavigated the world in 1580 in a ship named *Golden Hind*. This belief inspired Diteman to emulate his hero by crossing the Atlantic in an aircraft.

Diteman had purchased a small Barling monoplane in 1928 during a cattle-selling trip in Kansas City, Missouri. Although the aircraft was not designed for long-distance travel, Diteman believed it had the capability of flying

"Sandy" Wynne-Eyton almost died when his de Havilland Puss Moth (above and below) caught fire while taking off from St. John's.

non-stop to England from Newfoundland. While privately preparing for a transatlantic flight, Diteman explained, to anybody who asked, that he had acquired the plane for a tour of South America. Prior to departing for Newfoundland from his home in Billings, Montana, he informed friends that he was flying to Saint John, New Brunswick, to conduct genealogical research on his family.

In October 1929 Diteman flew to Roosevelt Field, and from there to Fredericton, New Brunswick. The airman arrived in St. John's, Newfoundland, on October 16. At Lester's Field, where Alcock and Brown had departed for Ireland in 1919, Mayor Tasker Cook attended the christening of Diteman's plane, which he named *Golden Hind*.

Prior to leaving Newfoundland for England, Diteman wrote a letter that offered a glimpse into his fragile mental state:

> Drake did not bring me here nor to London – albeit I am a descendant. Sorry if I hurt any one's feelings and I meant no harm.
> Dite
> P.S. Many thanks, you will hear from me.

After spending three days in St.

An atempted round-the-world voyage by John Mears (above and right) ended soon after arriving in Harbour Grace from Roosevelt Field, Long Island. The aircraft crashed during takeoff and was damaged beyond repair.

John's, Diteman flew to Harbour Grace. He took off for England on October 22, and was never seen again.

Eight months later, in June 1930, the dense fog that frequently blankets much of the northwest Atlantic almost claimed the life of Australia's greatest flying hero, Charles Kingsford-Smith, who was completing a round-the-world flight. If not for the experience and caution of Kingsford-Smith and his crew, the *Southern Cross* would have disappeared – as so many others had – flying west across the North Atlantic.

By this time only the *Bremen* had completed an east-to-west crossing of the Atlantic, and this effort could only have been termed a total success since it had ended in a crash-landing on Greenly Island.

Two years after the *Bremen* crash, Kingsford-Smith decided to succeed where the German flight had failed, by achieving the first non-stop flight to New York from Europe.

Kingsford-Smith, who had spent four years of his childhood in Vancouver, British Columbia, flew fighter planes for Australia during the First World War. After the war he earned a living as a barnstormer and Hollywood stunt pilot. By 1930 Kingsford-Smith was already well known for his endurance flights. In 1928 he and a crew of three had achieved the first flight across the Pacific, from California to Australia with refuelling stops in Hawaii and Fiji. The Pacific crossing was made in the *Southern Cross*, a tri-motor Fokker aircraft. Kingsford-Smith followed this achievement with the first non-stop flight across Australia.

By 1930 Kingsford-Smith was ready to attempt a round-the-world flight, and in May the *Southern Cross* took off from Oakland, California. The plane flew westward across the Pacific and reached Baldonnel Airfield in Dublin, Ireland, after three weeks of travel.

For the dangerous Atlantic crossing Kingsford-Smith selected as his crew people who were familiar with flying in poor weather conditions over the world's oceans. Joining him were Dutch KLM pilot Evert van Dyk, navigator Jonathon Saul of Ireland, and New Zealand radio operator John Stannage.

The Irish government went to extraordinary lengths to assist the *Southern Cross*. It ordered 500 soldiers from the Curragh Camp military training centre to prepare a makeshift airstrip on a beach at Portmarnock, a community northeast of Dublin.

Additional fuel tanks were installed inside the cabin of the plane, providing enough gasoline to last 35 hours of flying. The plane was flown to Portmarnock, where it waited for almost three weeks for weather to clear over the North Atlantic.

Before leaving Ireland on the Atlantic crossing, Kingsford-Smith was presented with a golden shamrock, for luck. The *Southern Cross* and its crew would need all the luck – and extra gasoline – they could muster.

The *Southern Cross* finally took off on June 23. For provisions, each man was given a thermos of coffee; ham, egg, and tomato sandwiches; and malted milk tablets. The plane made steady progress until nightfall, when it encountered a massive fog bank. Kingsford-Smith and van Dyk tried to escape it by climbing and descending, but to no avail. When the radio ceased working, and then blinked back on, Kingsford-Smith surmised it happened because the antenna had struck the ocean waves.

Despite carrying a radio, the flyers were unable to get their bearings, since navigation still relied heavily upon visual cues from the stars, moon, and sun. The *Southern Cross* was speeding blindly through the night. The crew became increasingly frustrated, especially with the lack of communication from Newfoundland's Cape Race radio station. Kingsford-Smith's log entry at 11:02 p.m. on June 24 reads: "We're within 100 miles [160 kilometres] of Cape Race, but the big stiff doesn't reply." The *Southern Cross* was, in fact, at least 240 kilometres away at the time. Fatigue also became a hazard, as another log entry shows: "Dickens of a struggle to stay awake now. Drone makes you tired."

As dawn broke, the Cape Race and Belle Isle radio stations guided the plane towards Harbour Grace. By now the men were becoming desperate, and Kingsford-Smith pleaded for a plane to be sent that could guide them through the fog to the airstrip.

No help came. Eventually, however, a hole appeared in the fog over Harbour Grace. The *Southern Cross* was finally able to land after being airborne for over 31 hours. Even though the Harbour Grace stop was unplanned, the *Southern Cross* had become the first aircraft to arrive safely in North America after flying the treacherous east-west route across the Atlantic.

They left Harbour Grace the next day and landed at Roosevelt Field, Long Island, a few hours later. Kingsford-Smith had failed to fly non-stop from Europe to New York, an achievement that would elude aviators until 1932. The *Southern Cross* then embarked on an air tour of the United States before completing its round-the-world journey at Oakland on July 4 – Independence Day.

Like many of the early aviators, Kingsford-Smith died in an air accident; he disappeared over the Bay of Bengal in 1935 when his aircraft flew into a typhoon.

As Kingsford-Smith was preparing for his transatlantic flight in Ireland, Dublin's Baldonnel Airfield had hosted another flyer who was busy rethinking his decision to fly the dangerous westward route across the ocean.

Charles Sanford Wynne-Eyton had won Britain's Distinguished Service Order as a pilot during the First World War and continued to serve with the Royal Air Force until 1926. He then found work in Rhodesia, where he ran a tobacco plantation.

In June 1930 he arrived in Dublin with his small, single-engined de Havilland Puss Moth monoplane. He planned to fly the machine to Newfoundland. The plane, which had been modified by adding fuel tanks, was nicknamed "the flying gas tank." Wynne-Eyton claimed that this extra fuel capacity would give the plane a range of 4,800 kilometres. He planned to release one of the tanks, which was situated underneath the plane, in flight once it had been emptied.

After assessing the risks of flying westward across the Atlantic, Wynne-Eyton decided to complete the less risky Newfoundland-to-Ireland route instead. He crated the plane and travelled by ship to St. John's. From St. John's he planned to fly to Harbour Grace, and from there to Europe.

On July 6 Wynne-Eyton took off from Lester's Field, bound for Harbour Grace. But disaster struck as the plane left the airstrip. Its engine stalled only twelve metres from the ground. The Puss Moth dived, bursting into flames as it crashed to earth.

Spectators rushed to the burning plane and managed to pull its unconscious pilot from the cockpit. Wynne-Eyton survived, but spent several weeks in hospital recovering from severe burns. He returned home and did not attempt another transatlantic flight. Wynne-Eyton flew aircraft for Britain during the Second World War. He was killed in an air crash in northern France in 1944.

Show business melded easily with aviation in the post-First World War years, an era when record-setting pilots could expect tickertape parades in New York City and enthusiastic crowds during victory tours of North America and Europe. Ruth Elder had used her unsuccessful transatlantic flight to land acting roles in feature films, while others, such as Lindbergh, Earhart, and Byrd, established lucrative careers as aviators due to their exploits.

It was therefore fitting that a Broadway producer would play a prominent role in the glamorous world of long-distance aviation. In addition to being a highly acclaimed member of New York's theatrical fraternity, John Mears had previously set records for travelling around the world.

In 1913, Mears circled the globe in 35 days by ship, train, and flying boat. His record stood for thirteen years, until broken in 1926 by Linton Wells and Edward Evans. Like Mears, Wells and Evans used several modes of transportation.

Mears was determined to set the record once again. In 1928 he hired pilot Charles Collyer to fly a Fairchild FC-2W monoplane, named *City of New York*, on a highly publicized voyage around the world. The round-the-world trip, dubbed "Racing the Moon" by Mears, was supposed to be completed in under 27 days, the time it takes for the moon to orbit the earth. The two men, and a dog named Tailwind, succeeded, and circled the world in a record-setting 23 days. Although impressive for the time, Mears and Collyer did not travel entirely by plane. To cross the Atlantic Ocean, they had boarded a ship.

Two years later, in 1930, Mears decided to circumnavigate the world entirely by plane, without resorting to ship travel, and to beat the time of 21 days set by an airship, the *Graf Zeppelin*, in 1929. He purchased a Lockheed Vega, and again named it *City of New York*. The Vega, like the Fairchild used in 1928, was christened by the wife of New York City's mayor, Jimmy Walker.

Mears needed a new dog as mascot for the journey because Tailwind had been killed by a hit-and-run driver in 1929. No ordinary dog would do, however, and Mears successfully appealed to Canadian-born screen actress Mary Pickford, who gave him her dog, which he named Tailwind II. Numerous delays compelled Mears's pilot to resign, but the *City of New York* finally departed Roosevelt Field on August 2, 1930, with a replacement pilot, Henry Brown.

The plane landed at Harbour Grace after an eight-hour flight.

Erroll Boyd (left) became the first Canadian to complete a transatlantic flight. His navigator was an American, Harry Connor (right).

Early next morning, Mears and Brown prepared for takeoff in the darkness using flares and lights to guide the *City of New York* down the airstrip. As the plane gained speed, its wheels struck a rock, causing it to careen off the runway and into the brush. The men escaped unharmed, but the plane was extensively damaged.

Pickford's dog disappeared in the mishap. Mears darkly surmised that a local resident had kidnapped Tailwind II, and he offered a reward for the dog's return. The animal was found two days after the accident, filthy, hungry, and drenched by rain. Mears returned to New York. He made no further attempts at flying around the world.

Two months after Mears's accident, Harbour Grace hosted the man who became Canada's first transatlantic flying hero.

Erroll Boyd was an unlikely candidate to set the Canadian record because the Toronto native had been denied entry as a member of Britain's new Royal Flying Corps during the initial months of the First World War,

N-X-237
COLUMBIA
1597

due to colour-blindness. Persistence paid off, however, as the Royal Naval Air Service, which used a different visual test, accepted Boyd as a pilot. He was taught to fly by John Alcock, who would make the first non-stop transatlantic flight in 1919.

In 1915, Boyd's wartime activities came to an abrupt end, as his aircraft was shot down over the Netherlands, which was then neutral. Although he was officially designated a prisoner of war, Boyd was allowed to travel freely within the country and permitted visits to the United States – which was also neutral until 1917 – on medical grounds.

While in New York he married a Broadway dancer, Evelyn Carberry, whom he had met in Toronto prior to the war. Through her he met singer Al Jolson and other entertainers, encounters that inspired Boyd to put his talents as a songwriter to use. He wrote several popular songs, one of which became a hit on Broadway after the war.

Despite his wartime success as a songwriter and a pilot, neither occupation earned Boyd a steady income; during the immediate post-war years he worked at a variety of jobs, including at a car rental agency and a hotel.

Columbia was renamed *Maple Leaf* by Erroll Boyd for his transatlantic flight. It set out from Harbour Grace on October 9, 1930.

During the mid-1920s, he obtained employment as a charter and airmail pilot and gained considerable experience flying long distances and coping with harsh weather conditions. After being hired by Charles Levine's Columbia Aircraft Company in 1929, Boyd and navigator Harry Connor flew the *Columbia* on a record-breaking non-stop return flight between New York and Bermuda in June 1930. This was the same aircraft that Levine and Clarence Chamberlin had flown across the Atlantic three years earlier. In 1928, Mabel Boll flew the *Columbia* to Newfoundland on her aborted bid to become the first woman to fly across the Atlantic.

Connor's navigational skills so impressed Boyd during their Bermuda flight that he invited the American to accompany him on his transatlantic flight later in 1930. But first, Boyd needed to get the *Columbia* to Harbour Grace – this turned into a trial equivalent in difficulty to the transatlantic flight itself. With little money of his own, Boyd relied upon sponsors to finance the project: during the Great Depression such support was not readily available.

Finding no backers in Toronto, Boyd sought help in Montreal. However, upon landing at St. Hubert Airport on September 1, he was met by the Royal Canadian Mounted Police. Officers impounded the plane – which Boyd had renamed the *Maple Leaf* – on behalf of one of Levine's creditors. Although he succeeded in finding financial support for the transatlantic flight, Boyd's hotel bills remained unpaid and continued to mount as the days passed.

The court finally released the plane after almost two weeks, but by then neither Boyd nor Connor had any money to pay the hotel. Rather than wait for a benefactor, the two men sneaked out of the building and flew to Prince Edward Island. A brewery eventually paid the aviators' Montreal hotel charges, and on September 23, after ten days on PEI, the *Maple Leaf* finally took off and landed in Harbour Grace.

By this time autumn had set in, and with it the onset of unpredictable weather. Although nobody had successfully crossed the Atlantic so late in the year, Boyd and Connor decided to proceed with their flight to Croydon Airport in London.

On October 9, after waiting over two weeks for favourable weather, Boyd and Connor took off from Harbour Grace. Like the *City of New York* a few months earlier, the *Maple Leaf* almost ended its journey in a heap of tangled metal – it skidded along the airstrip and came close to flipping over. Once the plane was airborne, the airmen battled strong winds, and at higher altitudes ice built up on the wings.

Their closest brush with danger occurred the next day as the *Maple Leaf* neared Europe; a clogged line was preventing them from using fuel in the reserve tank. With almost no usable fuel remaining, Boyd and Connor searched the horizon for land. Luck was with them, as they soon came upon the Isles of Scilly, a group of islands 45 kilometres southwest of Land's End, the most westerly point in mainland England.

Boyd landed the *Maple Leaf* at a beach on the island of Tresco, which had been used by Britain during the First World War as a seaplane base.

With this achievement, the *Maple Leaf* became the first plane to fly across the Atlantic outside the summer months, and Boyd the first Canadian to complete a transatlantic flight. The aircraft also became the first to cross the Atlantic twice, the first time being in 1927.

Both Boyd and Connor continued flying throughout the 1930s, with Boyd working as an aviation columnist with the *Toronto Star Weekly*. During the Second World War, Boyd moved to the United States where he recruited pilots for Canada and Great Britain. In 1938, millionaire Howard Hughes hired Connor to navigate his plane on what became a record-setting round-the-world flight.

CHAPTER 5

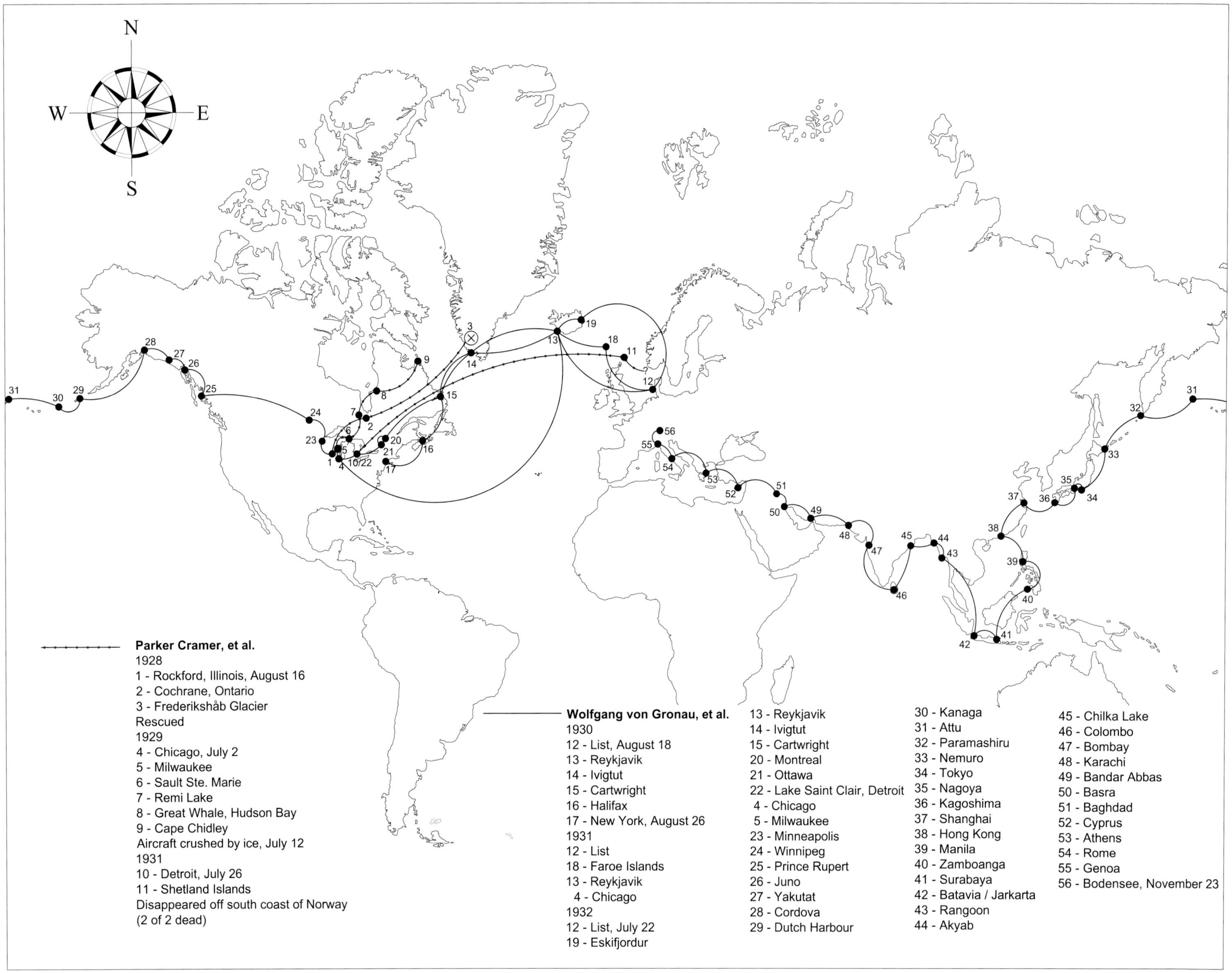
N
W
E
S
Parker Cramer, et al.
1928
1 - Rockford, Illinois, August 16
2 - Cochrane, Ontario
3 - Frederikshåb Glacier
Rescued
1929
4 - Chicago, July 2
5 - Milwaukee
6 - Sault Ste. Marie
7 - Remi Lake
8 - Great Whale, Hudson Bay
9 - Cape Chidley
Aircraft crushed by ice, July 12
1931
10 - Detroit, July 26
11 - Shetland Islands
Disappeared off south coast of Norway
(2 of 2 dead)
Wolfgang von Gronau, et al.
1930
12 - List, August 18
13 - Reykjavik
14 - Ivigtut
15 - Cartwright
16 - Halifax
17 - New York, August 26
1931
12 - List
18 - Faroe Islands
13 - Reykjavik
4 - Chicago
1932
12 - List, July 22
19 - Eskifjordur
13 - Reykjavik
14 - Ivigtut
15 - Cartwright
20 - Montreal
21 - Ottawa
22 - Lake Saint Clair, Detroit
4 - Chicago
5 - Milwaukee
23 - Minneapolis
24 - Winnipeg
25 - Prince Rupert
26 - Juno
27 - Yakutat
28 - Cordova
29 - Dutch Harbour
30 - Kanaga
31 - Attu
32 - Paramashiru
33 - Nemuro
34 - Tokyo
35 - Nagoya
36 - Kagoshima
37 - Shanghai
38 - Hong Kong
39 - Manila
40 - Zamboanga
41 - Surabaya
42 - Batavia / Jarkarta
43 - Rangoon
44 - Akyab
45 - Chilka Lake
46 - Colombo
47 - Bombay
48 - Karachi
49 - Bandar Abbas
50 - Basra
51 - Baghdad
52 - Cyprus
53 - Athens
54 - Rome
55 - Genoa
56 - Bodensee, November 23

FLYING THE ARCTIC ROUTE

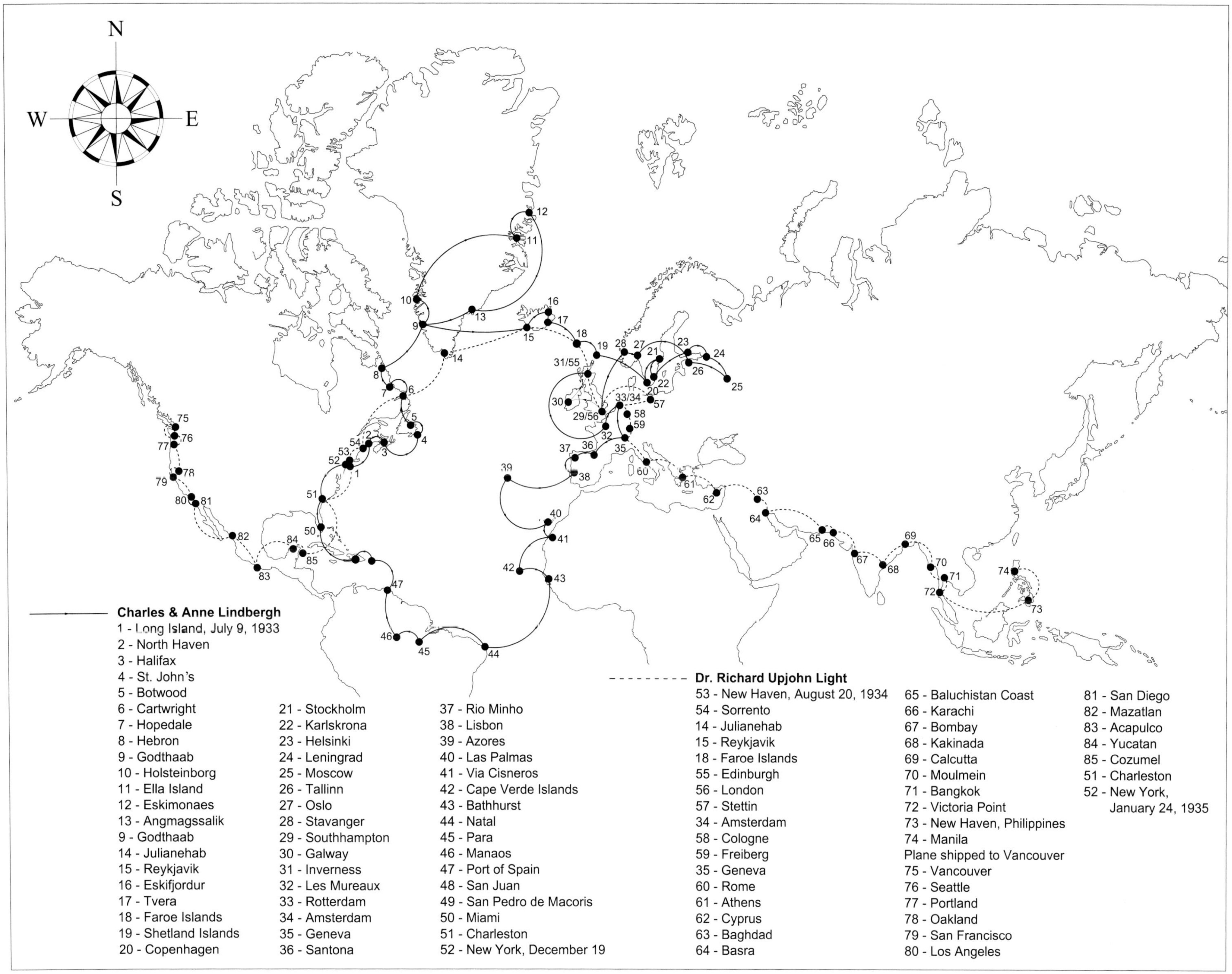

As the number of deaths and accidents among Atlantic flyers grew during the late 1920s, many aviators sought ways of reducing the risks. Charles Lindbergh had mused about converting aging freighters into permanent refuelling bases at sea. The American Navy was building its first fleet of aircraft carriers at the time, but the costs of maintaining similar ships for civilian, commercial use made the idea impractical.

Others believed the answer lay in avoiding the foggy, windswept North Atlantic altogether. Under this proposal, new routes would be charted for the coming age of commercial airliners, with planes flying over Greenland and Canada's far north. Roald Amundsen had already crossed over the North Pole in 1926, flying an airship to Alaska from the Norwegian island of Spitsbergen. For two men, Parker Dresser Cramer of the United States and Wolfgang von Gronau of Germany, Amundsen's flight suggested the viability of commercial airline routes using the Arctic to link Europe with North America.

The crew of a German seaplane, led by pilot Wolfgang von Gronau, paid a visit to the Labrador community of Cartwright in 1930 (above) after crossing the Atlantic from Europe.

Cramer, whose uncle established the Dresser manufacturing firm, would make three attempts at flying over the Far North to Europe from Chicago. Each venture ended in disaster, thereby buttressing the impression that Arctic flying was at least as dangerous as crossing the fog-bound North Atlantic.

In his first attempt, in 1928, Cramer and another airman, Bert Hassell, embarked on a flight from Illinois to Stockholm, Sweden. They became lost while flying over Greenland. As the plane ran out of fuel, Cramer was forced to land on Frederikshåb Glacier in southwestern Greenland, several hundred kilometres off course.

Carrying a gun and some food, the men hiked towards a meteorological station, which they thought could be reached in two days. They seriously miscalculated; Cramer and Hassell were finally rescued by an Inuit hunter after walking for fifteen days across Greenland.

A year later, in July 1929, Cramer made a second attempt, this time with the support of Robert McCormick, the eccentric publisher of the Chicago *Tribune*. McCormick had created a sensation in journalism by ordering his reporters to abandon traditional spelling and to write phonetically – an innovation he soon abandoned. McCormick owned two Sikorsky S-38 amphibious aircraft, one of which was named *Untin Bowler*, a slang term for a type of English hunting hat.

On July 3 Cramer, along with co-pilot Robert Gast and *Tribune* aviation editor Robert Wood, took off from Chicago in the *Untin Bowler*. They planned to fly over the Arctic and land in Berlin. The flyers would then turn around and travel the same route back to Chicago. But poor weather compelled them to take a much longer tour of Canada's Far North than anticipated. They spent five days battling strong winds and dense fog, and were forced to refuel at isolated government outposts along James Bay.

Their goal was to land at the Royal Canadian Mounted Police outpost of Port Burwell, located near Cape Chidley on Killinek Island at the northernmost tip of

Labrador, before heading east across the Greenland ice cap. The radio operator at Port Burwell would be critical in guiding the *Untin Bowler* across Canada's North.

Throughout the journey, Wood had been transmitting news articles from the plane back to the *Tribune* in Chicago. His reports illustrated the difficulties faced by Cramer and Gast:

> Fog lay in about the hills as the *Bowler* flew northward. We caught the first sight of the lofty Labrador mountains to the east. Cape Chidley and the whole northern point of Labrador was blanketed in fog, with only the tops of the mountains and hills visible.

In another report Wood described the isolation and rugged beauty of Ungava Bay:

> It is a veritable "no man's land" of muskeg swamps, rock, and stunted forests. Of government patrols in this territory there are none – no fire rangers and no forest fire protective planes, for it is all beyond the limit of marketable timber growth, and stunted birch, balsam fir, black and white spruce, and larch, dot the landscape.

On July 9 the *Untin Bowler* finally landed at Amittoq Inlet near Port Burwell. Local Inuit men were hired to carry drums of fuel to the aircraft, but poor weather repeatedly prevented it from taking off. Worse yet, sea ice began moving into the inlet. The plane taxied into nearby Fox Harbour, where it was moored beneath an abandoned Moravian church. Still, slabs of ice threatened the *Untin Bowler*. The crew and volunteers managed to push ice away several times, but on July 12 nature finally prevailed. They could only watch as their plane crumpled and sank beneath a large slab of ice. The trio were forced to remain at Port Burwell until a supply ship arrived at the end of July.

Cramer's two failed attempts did not dissuade German aviator Wolfgang von Gronau, who also believed that it was safer and more economically feasible to establish commercial routes over the Arctic than to fly across the Atlantic. Despite being one of Germany's most influential airmen, Gronau had trouble convincing government authorities of the merits of using flying boats to carry mail and passengers along his proposed Greenland route.

The son of an army general, Gronau had flown seaplanes during the First World War. He became a director of a government-financed flying school in the mid-1920s, teaching students to operate float planes and flying boats. To train his

An attempt by Parker Cramer (right) to reach Europe by crossing over the Arctic ended prematurely when pack ice crushed his plane at Port Burwell (above), on Killinek Island at the northernmost tip of Labrador, in 1929.

students, Gronau often flew with them to Iceland in a Dornier Wal flying boat, the D-1422. This aircraft had been flown by Roald Amundsen in an unsuccessful attempt at crossing the North Pole in 1925.

On August 18, 1930, the Wal – German for whale – took off from the port of List on the North Sea on what Gronau said was a routine flight from Germany to Iceland. In fact, he intended to fly to New York City. Gronau was highly secretive about his plans. "I did not want blame attached to my name as a pilot if originally announced plans were not carried out successfully," explained Gronau, who did not even confide in his wife. Accompanying Gronau was co-pilot Edouard Zimmer, a recent graduate of the flying school, as well as mechanic Franz Hack and radio operator Fritz Albrecht.

The Wal completed the journey with relative ease. After refuelling at Iceland and Greenland, the plane arrived in Cartwright, Labrador, on August 23, where the crew stayed overnight. The next day they flew to Halifax, Nova Scotia. On August 26 Gronau and his crew were welcomed by thousands of onlookers as they landed under the Statue of Liberty in New York harbour.

Despite Gronau's success, air-

Von Gronau and three crewmen flew their Dornier DOJ11b Wal from Germany to Cartwright to test the viability of using northern routes across the Atlantic Ocean. The seaplane flew on to Quebec, Western Canada, and the United States.

lines remained skeptical about flying the so-called Greenland route. So in August 1931 both Gronau and Cramer decided to fly once again across the Arctic and northern Europe. Cramer took off first, leaving Detroit with the goal of landing in Copenhagen, Denmark. On this, his third, attempt, Cramer came close to success. This time, he and navigator Oliver Paquette had flown over northern Canada and Greenland without incident and landed safely at the Shetland Islands. Then, after taking off on August 9 for the last leg of their trip, their aircraft disappeared as it came within sight of Norway. A month later, wreckage from the plane was found in the North Sea, and five months after the crash Cramer's briefcase containing letters and maps were found by the crew of a fishing boat near the Norwegian city of Stavanger.

A further blow to Arctic flight came from Gronau, whose experiences flying over Greenland proved so harrowing that he changed his mind about the viability of the northern route. Less than a week after Cramer's disappearance, Gronau and his crew arrived safely in Chicago, following a flight from Germany. It was a perilous journey, as the Wal came close to crashing into Greenland's ice cap. Due to what Gronau said was the "rarified atmosphere" over Greenland and the towering ice fields, the crew had to jettison fuel in order to gain altitude and

Enthusiasm for Parker Cramer's attempt at reaching Europe by flying over the Arctic was reflected in media coverage, including this lighthearted cartoon in the ***Chicago Tribune***.

Charles and Anne Lindbergh were greeted by dignitaries upon their arrival at Bay Bulls Big Pond, near St. John's. They flew a seaplane named *Tingmissartoq*, a Greenland Inuit word meaning "one who flies like a bird."

avoid disaster.

Despite such misgivings, Gronau again flew over Greenland the following year when he completed a round-the-world flight in his aging aircraft. The plane, with co-pilot Geth von Roth, Albrecht, and Hack, had left Germany on July 22, 1932. They stopped at Iceland and Greenland, before arriving at Cartwright on July 26. The D-1422 then flew west, touching down at Montreal, Ottawa, Detroit, Chicago, Milwaukee, Winnipeg, Prince Rupert, and three communities in Alaska. The plane proceeded to the Soviet Union, southeast Asia, the Middle East, and Italy. Four months after starting the journey, on November 23, the plane returned to Germany.

Although Gronau had circumnavigated the globe, German authorities remained skeptical about long-distance travel using heavier-than-air machines. The country had become the world's leader in airship travel and dirigibles were still believed to be the only practical way for passengers to fly over the Atlantic. Gronau's dream for a northern route failed to come to fruition, but his career soared after the Nazis took power in 1932. He became president of the German Aero Club, and in 1939 was appointed air attaché to the German embassy in Tokyo.

In July 1933 North American newspapers paid scant attention to the activities of the world's best-known aviator, Charles Lindbergh, when he and his wife, Anne, stepped from their plane at a lake near St. John's to shake hands with a group of airline representatives and government officials.

Public attention was focused on a fleet of Italian seaplanes, led by Italo Balbo, which was in Labrador at the time, en route from Europe to Chicago to celebrate the tenth anniversary of Benito Mussolini's regime. While the Italians were creating headlines, the Lindberghs and leading members of the airline industry were discussing crucial issues that would eventually result in regular transatlantic passenger flights between Europe and North America.

Pan American World Airways had hired Lindbergh and his wife, Anne Morrow Lindbergh, to determine the viability of using the Greenland route to cross the Atlantic. In conjunction with a research ship, the *Jellinge*, they would survey possible seaplane landing sites in Newfoundland, Labrador, Greenland, and Iceland.

The timing of the survey was no coincidence – it began at the commencement of an international meeting in St. John's involving Pan American, British-based Imperial Airways, and government officials from United States, Great Britain, Canada, and Newfoundland. When deliberations ended two weeks later, the participants had agreed to cooper-

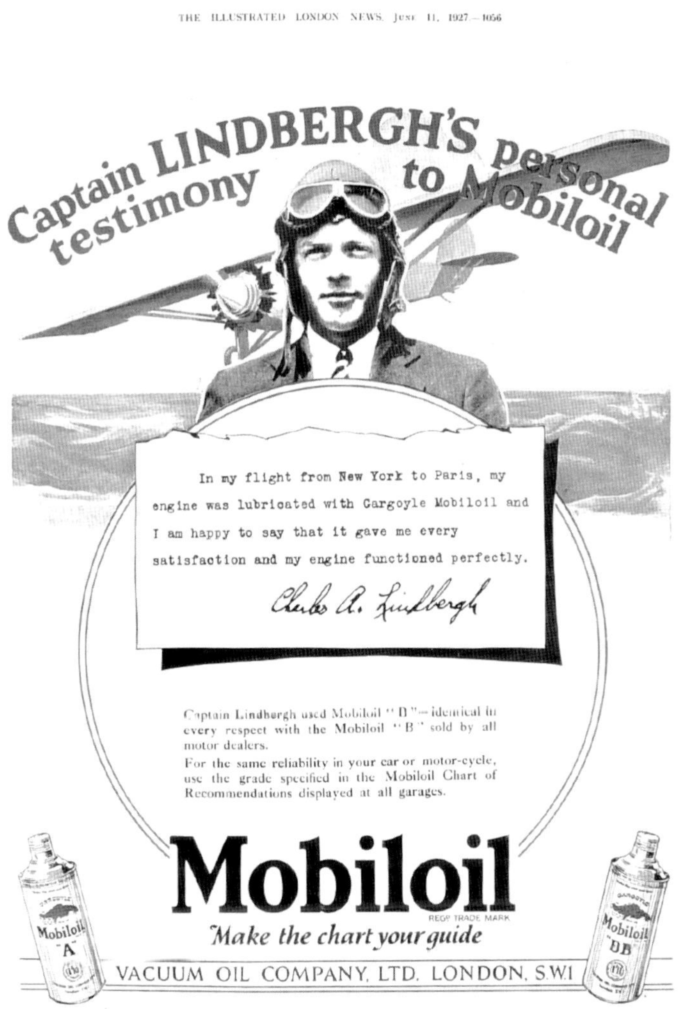

In 1932, the kidnapping and murder of Charles and Anne Lindbergh's son, Charles Jr., transfixed the public and led to the most extensive manhunt in US history.

Their twenty-month-old son was plucked from his crib, with the kidnapper leaving a ransom note demanding $50,000 for his safe return. The Lindberghs paid the ransom, but the boy's badly decomposed body was later recovered near their New Jersey estate.

Two years later, in 1934, police arrested a German immigrant, Bruno Hauptmann. The trial, dubbed by author H.L. Mencken as "The greatest story since the Resurrection," resulted in Hauptmann's conviction. He was executed by electric chair in 1936.

ate on establishing a transatlantic air service.

On July 9 the Lindberghs took off from Long Island in a Lockheed Sirius aircraft. The plane was named *Tingmissartoq*, a Greenland Inuit word meaning "one who flies like a bird." With Charles in the pilot seat and Anne as navigator and radio operator, the pair stopped in Maine and Halifax before arriving in Newfoundland on July 12. Their scheduled landing spot was Bay Bulls Big Pond, twelve miles south of St. John's. Instead they mistakenly landed at the city on Quidi Vidi Lake. A group of surprised onlookers gave the unexpected visitors directions, and within minutes the Lindberghs arrived at their destination.

Pan American's representative at the international meeting, Alan Wilson, greeted the aviators as they stepped ashore. The Lindberghs stayed at the home of businessman Leonard Outerbridge, who would later serve as Newfoundland's first lieutenant-governor when it became a province of Canada in 1949. Fêted at a party held in their honour at the Hotel Newfoundland by delegates to the aviation conference, the Lindberghs socialized for only an hour before retiring to the privacy of the Outerbridge home. This reluctance to appear at public events would be repeated frequently over the course of their travels that year. Just one year earlier, their son, Charles Jr., had been kidnapped and murdered. In the months and years following this tragedy, the couple would become increasingly reclusive.

On July 14 they took off from Bay Bulls Big Pond and headed west to Botwood, a community in Notre Dame Bay in central Newfoundland. Botwood was chosen because it had already been used as a float plane base by Newfoundland aviator Sidney Cotton, who began an airmail and aerial survey company in 1921. With Botwood as the base for the Aerial Survey Company, Cotton delivered mail to isolated communities throughout the island and Labrador.

Although the Lindberghs stayed only a few hours in Botwood, their visit would prove pivotal to the town's future. Its deep harbour and location on a great circle route between Europe and North America would lead to Botwood becoming a refuelling stop for transatlantic passenger planes within six years of the Lindberghs' arrival.

Their next stop was Cartwright, in Labrador, where they had hoped to meet Balbo and his fleet of seaplanes. But the Italians had left one day earlier, on July 13, for New Brunswick. Dense fog then

Charles and Anne Lindbergh at Roosevelt Airfield, New York, in December 1929.

grounded the *Tingmissartoq* for almost a week.

While the couple waited in Cartwright, an agreement was being signed in St. John's that would allow two of the world's largest airlines to begin commercial transatlantic air service. Imperial Airways signed an agreement with the government of Newfoundland, giving it a licence to fly aircraft between the island and Great Britain. Although the licence gave Imperial exclusive rights to this service, the airline had established another agreement with Pan American which would result in both airlines sharing the Botwood facilities. In 1933, the two companies began investing heavily in new fleets of flying boats with the capacity to fly across the Pacific and the Atlantic oceans.

On July 20, the skies finally cleared over Cartwright, allowing the Lindberghs to fly north towards Greenland. They spent a night in the Labrador village of Hopedale, and then stopped for a few hours at the Moravian missionary community of Hebron, located close to the northernmost tip of Labrador.

The *Tingmissartoq* finally joined the *Jellinge* at Godthaab in western Greenland on July 22, following a six-hour flight from Hebron. A group of Inuit and Danish officials cheered the plane from atop cliffs overlooking the town. Children presented Anne Lindbergh with a bouquet of Greenland poppies as she stepped ashore. At an official reception that evening she told Danish officials:

> I shall never forget the majestic beauty of the mountains as we first saw them towering, glistening white, above the clouds. We seemed to be in a new and fantastic world.

During their 24 days in Greenland, the Lindberghs surveyed coastal areas of the Danish territory, flying from Canada's Baffin Island to Greenland's east coast. After completing this portion of their work, the Lindberghs flew to Iceland. They finished the arctic survey at the Faroe Islands in mid-August.

With the formal work now complete, the Lindberghs spent the remainder of the year travelling, ostensibly on vacation. Huge crowds met them in Denmark, Sweden, Norway, the Soviet Union, Netherlands, France, Spain, and Portugal. In late November, the famous couple travelled to the resort city of Villa Cisneros (renamed Dahkla) in Western Sahara. Two weeks later, they flew across the southern Atlantic, reaching the Brazilian mainland on December 8. Their 21-nation tour ended in Miami, Florida – where Pan American was then headquartered.

Although the northern air survey would prove invaluable to air transportation during the Second World War, Charles Lindbergh's findings ended the Greenland route as a viable option for commercial seaplane travel. In his report, Lindbergh wrote:

> [The survey] left no doubt in my mind about the wisdom of establishing the first air route to Europe by way of Bermuda and the Azores. Ice conditions along a more northern route would prevent operation of flying boats [during the winter] ... we could never give satisfac-

tory service to transatlantic passengers if we had to move our terminal with the seasons.

With the completion of the Lindberghs' mission, few aviators would attempt to fly over Greenland until the outbreak of the Second World War. A notable exception was physician and world traveller, Richard Upjohn Light.

When Light landed his plane at Cartwright, Labrador, on August 23, 1934, he had hoped to tour the medical facilities run by famed humanitarian Sir Wilfred Grenfell; but Labrador's only hospital had been destroyed by fire a few days prior to Light's arrival. He witnessed medical staff building makeshift operating tables out of crates left by Italo Balbo's air squadron a year earlier, and attended the funeral for a girl who perished in the blaze.

Light, who had studied neurosurgery under the tutelage of Harvey Cushing, the world's leading expert in the field, had already commenced a promising career as a physician and an instructor in surgery at Yale University. Nevertheless, he was determined to travel around the world. He had learned to fly in the late 1920s while serving in the US Army Air Corps. In 1934, he decided that he had built up enough experience as a pilot to fly around the world. Unlike many world flyers, Light was not concerned with breaking records. His goals were to experience other cultures and to tour areas of the planet rarely visited by Americans.

Dr. Richard Upjohn Light visited Labrador during a round-the-world tour in 1934. The Wisconsin native is seen here with his Pitcairne Mailwing aircraft in 1932.

Light and co-pilot Robert Wilson began their world trip at New Haven, Connecticut, on August 20. One of their first stops was Cartwright, an important refuelling stop for planes commencing the Arctic route to Europe. During his five days in Labrador, Light travelled by boat with a Grenfell Mission doctor to tiny communities along the coast.

After leaving Labrador, Light and Wilson stopped in Greenland and Iceland before arriving in Europe. The flyers reached the Philippines in December, but from there they decided not to risk a dangerous flight across the Pacific Ocean. Instead, they travelled by ship to Vancouver, British Columbia.

Light and Wilson visited Mexico and Cuba, and finished their five-month-long journey at New Haven on January 22, 1935. Two years later, Light would travel once more by air, this time crossing Africa with his first wife, Mary Meader. In recognition of his explorations, he was elected as president of the American Geographic Society from 1947 to 1956.

CHAPTER 6

Italy achieved a public relations victory by flying a squadron of aircraft to North America from Europe in 1933. Led by Air Minister Italo Balbo, the planes stopped in the Newfoundland community of Shoal Harbour in August that year.

FASCISM IN THE AIR

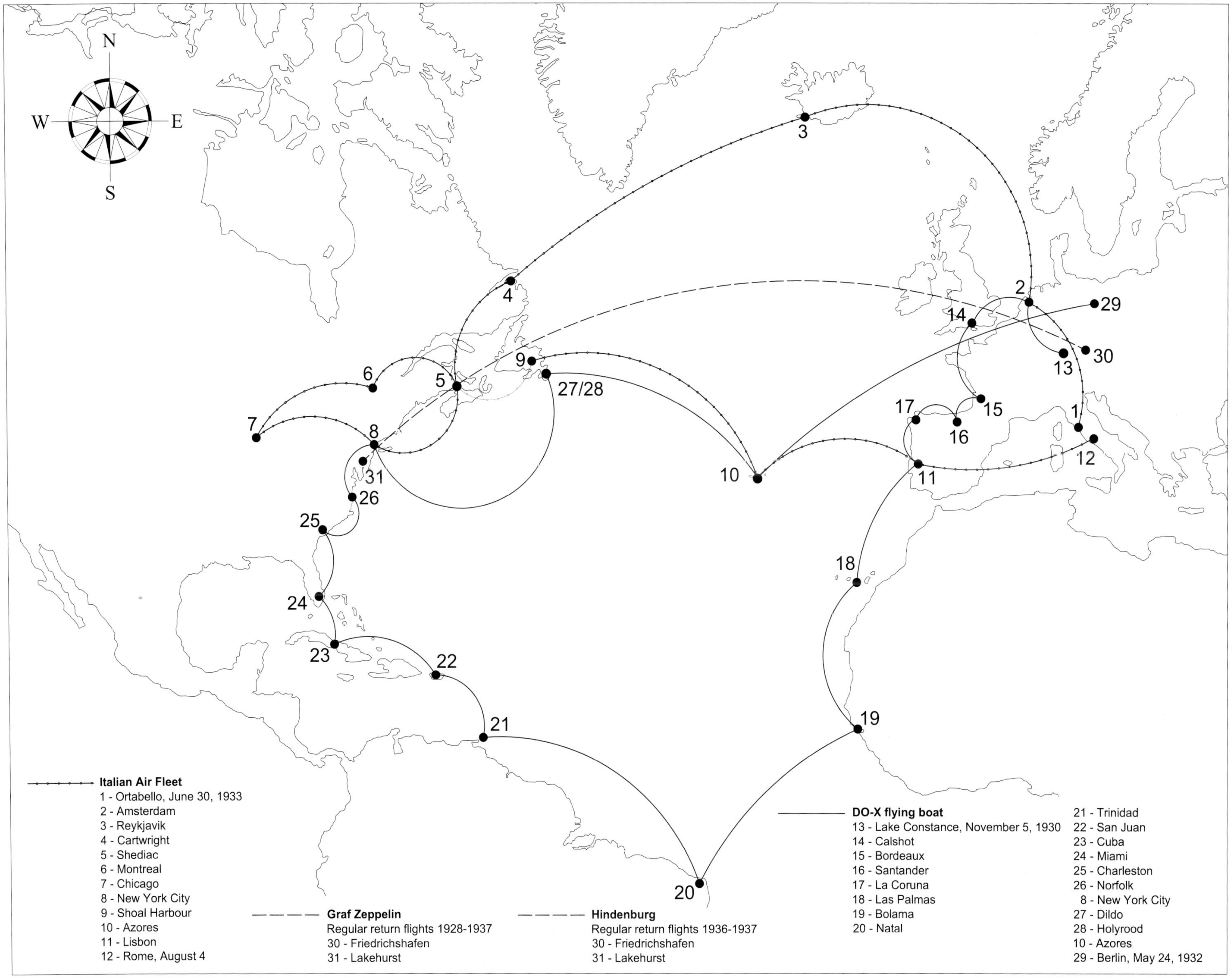

Public enthusiasm for aircraft – and the pioneers who set flying records – was reaching its zenith in the early 1930s, just as the Great Depression was sweeping across Europe and North America. Images of brave men and women soaring into the clouds charged the imaginations of people, millions of whom followed the exploits of their heroes in newspapers, on radio broadcasts, and in newsreels at movie theatres.

Such interest did not escape the notice of politicians, from small-town mayors to presidents, who clamoured to be photographed with record-setting pilots in hopes that they might benefit from reflected glory.

More importantly, aviation had become a useful public relations tool for many nations after the First World War, as governments linked triumph in the skies with national prowess. While the US, France, and Britain were among the leaders in the propaganda battle, the practice was elevated to new levels in the 1930s by Nazi Germany and Italy's fascist regime.

The most successful of these propagandists was Italian Air Minister Italo Balbo. As one of Benito Mussolini's most trusted aides, Balbo had been commander-in-chief of the brutal Black Shirt militias that brought the fascists to power in the early 1920s. Balbo ordered the beatings of communists, socialists, and trade unionists throughout Italy, and he was implicated in the murders of opposition figures.

Despite having no experience as an aviator, Balbo was appointed by Mussolini to a senior position in the air force in 1926. The debonair former bank clerk's leadership skills were readily apparent, and three years later he was appointed Air Minister. Balbo, who by then had earned his wings as a pilot, began reorganizing the air force into a propaganda tool for fascism.

The four-continent tour by Francesco de Pinedo in 1927 had already demonstrated the effectiveness of aviation as a force in uniting Italians. To reinforce the supremacy of fascism and of Italian air prowess, Balbo organized a spectacle never before attempted: a mass transatlantic flight by a squadron of aircraft.

On December 17, 1930, Balbo led twelve Savoia-Marchetti flying boats on a journey to Brazil. After taking off from Ortabello, Italy, the squadron spent several weeks in Portuguese Guinea (now Guinea Bissau). All of the aircraft reached Rio de Janiero safely on January 15, 1931.

The flight was a popular success. Huge crowds of spectators watched in fascination as the seaplanes roared overhead and landed one after another in the har-

The new breed of Italian and German dictatorships embraced the propaganda value of aviation. Italy's president, Benito Mussolini (centre right), ordered a former thug, Italo Balbo (at his side), to make the air force a showpiece of Italian power.

bour. When Balbo and his men flew back to Italy, a proud populace greeted them as if they were military heroes.

This proved to be a trial run for an even more ambitious project. In 1933 Balbo organized another mass flight across the Atlantic, but this one would involve twice the number of aircraft and require flying across the treacherous waters of the North Atlantic.

The spectacle was timed to coincide with two events: Mussolini's tenth year as president, and the Century of Progress Exposition in Chicago. A highlight of the world fair would be the unveiling, by Balbo, of a monument to Italian explorer Christopher Columbus.

Twenty-five newly built Savoia-Marchetti S-55 flying boats left Ortabello on June 30. The dozens of airmen chosen for the mission had spent several months training to fly in close formation. Each plane carried a four-man crew consisting of a pilot, co-pilot, engineer, and radio operator.

Balbo's squadron landed first at Amsterdam. There, one of the planes struck a dike while landing, drowning one crewman and damaging the aircraft beyond repair.

Despite the accident, the remaining aircraft flew on to Londonderry, Ireland. On July 5 the airmen took off for their next destination, Iceland. In Reykjavik, the arrival of a hundred Italian military men created a sensation among the populace; Balbo described their welcome as exceptionally warm and inviting. The Italians enjoyed a week touring Iceland, but seemed especially enthralled by the women they encountered on the island.

"My Atlantic lads, with the chivalry of their race, pay special attention to the Iceland girls, who are indeed very beautiful and charming," Balbo wrote in *My Air Armada*. "A few of the younger officers are inclined to fall seriously in love with some of the fair visitors to the hotel. But Iceland girls, simple and unsophisticated though they look, are too level-headed and calculating to pay serious attention to the romantic blarney of young Italians who have descended among them suddenly like a bolt from the blue, and who will vanish just as suddenly."

Indeed, on July 12, seven days after their arrival, the airmen took off and headed to their first North American stop, the community of Cartwright in Labrador.

Balbo described the hours spent in the ocean crossing as a period of constant anxiety, punctuated by moments of terror. For much of the journey, the squadron was obscured in rain clouds and dense banks of fog.

"At one point ... there is an absolute black out," Balbo recalled. "We can see nothing but the gleam of the instruments on the dashboard. We have run into a rain cloud and a torrential downpour drives right into our cabin. I relax my grip for a moment to write a note to Pezzani. But at this moment the plane heels. The

Balbo in Clarenville with John Puddester, Member of the House of Assembly, at his right. Puddester was later appointed to the Commission of Government that governed Newfoundland after its parliament was replaced by non-elected officials in 1934.

THE EVENING TELEGRAM

VOL. LVIII., NO. 75 FORTY PAGES TORONTO, SATURDAY, JULY 15, 1933 PRICE TWO CENTS

CITY AND SPORTS "THE PINK TELY"

ITALIAN FLYING FLEET NEARS CITY

HUGE DEMONSTRATION WELCOMES ITALIAN AERIAL ARMADA

TWO PLANES OFF ACROSS ATLANTIC GOALS DIFFERENT

Wiley Post After Record for Round-World Flight; Lithuanians Make Surprise Start Without Permission

Italian Planes Skirt Ontario's North Shore On Flight to Toronto

Balbo's Flying Fleet Logs 5,230 Miles

Delayed Start From Montreal Means Arrival in Queen City Some Time After 1.30

Evangelist to Fight But Asks Quiet Too

Leave Montreal Hour Before Noon

Mine Shares Jump Here To Near Record Heights Liquor Stocks Forge Up

T.T.C. Grants Free Transit To Sightless

Reports of Uncle Sam Favorably Studying Canadian Gold Mine Policy Boost Market

CHICAGO TERROR THREATENS FIFTY WITH ABDUCTION

Canadian-Born A. W. Cutten in Kidnappers' Book Revealed by Factor — Police Guard Homes

Thousands of Workers Get U.S. Wage Raises

Boy, 17, Fights 2 Men Robbing Father's Store

RUIN THREATENED BY CHINESE RIVERS

Great Britain To Play U.S. In Davis Cup

Dismissal of Bilbrough Ordered by School Board

plane bounces up and down with quick jerky movements. It is like an earthquake in the air, so to speak. I plead guilty to a fleeting spasm of fright."

The fog dissipated as Balbo's squadron reached Labrador, revealing the stark tree-lined coast and a glistening ocean dotted by patches of white pack ice. Balbo described his joy at seeing North America for the first time on his journey, which he celebrated by sending a telegraph message directly to Mussolini.

"A peculiar sense of semi-intoxication sweeps over me," he wrote. "The sea which sparkles beneath the rays of the sun seems to join in our exaltation with its crystalline foam-capped waves. Our Fascist native land may well be proud of its airmen. They have brought honour to the uniform of the black shirt, which they will wear proudly as they enter America."

The planes were met in Cartwright harbour by the *Alice*, an Italian support ship. A US seaplane, carrying a large contingent of journalists and photographers, had also arrived in advance of the squadron. The modest Labrador community seemed to have been an anticlimax after the excitement of crossing the Atlantic. Balbo described Cartwright as "a few little timber houses set against a rather depressing background of grayish-green vegetation. Thickets of stunted firs dotted among barren crags."

Residents of Cartwright cheerfully welcomed their Italian visitors. Many of them volunteered to help with a banquet held in honour of the airmen, while others assisted in refuelling the 24 flying boats.

Balbo was accorded a hero's welcome at home after a highly successful propaganda effort in North America.

Dorothy Fequet, one of the children who participated in the Cartwright festivities, recalled the events; her recollections were published in the September 1986 edition of Newfoundland and Labrador's *Them Days* magazine:

> Dormitories at the Grenfell Mission were requisi-

The Dornier flying boat of the early 1930s proved that the era of large passenger aircraft had not yet arrived. The DO-X, shown in Trinity Bay, could carry up to 70 people, but was badly underpowered despite carrying twelve engines. In 1934 the DO-X was placed in a German museum, but it was destroyed in a bombing raid on Berlin in 1945.

tioned by the *Alice* officers, to house the arriving aviators. Case after case of Italian wine came from the ship to load the Mission's cellars for a banquet planned for the airmen; this, to the chagrin of the Mission Matron, for cellars had never held anything stronger than soft drinks.

The banquet that night was something. Portraits of Mussolini and General Balbo adorned the walls with the modest portrait of Sir Wilfred Grenfell, the Mission founder. Italian cooks from the *Alice* prepared the food, while Mission waitresses and other volunteers rushed hither and yon, supplying the Italians with food and trying desperately to cope with the mounting flow of dirty dishes and utensils.

I can remember when Balbo's fleet came to Cartwright. Everybody that could was hired to put oil in the planes, and there was two children picked out to meet General Balbo at the Grenfell Mission dormitory. Noel Groves was dressed as an Italian boy – all dressed in Italian clothes – and I was dressed in Italian girl's clothes. We greeted him on the dormitory steps, shook hands with him and bowed to him.

The squadron took off the following morning, stopping next at Shediac, New Brunswick, where government officials greeted Balbo and his men. The next day, July 14, they flew along the St. Lawrence River and landed at Montreal.

There, a large, enthusiastic crowd met the airmen. An honour guard of Italian Canadians, wearing the Black Shirt uniforms of their fascist heroes, formed a cordon around the officers as they stepped ashore.

Even Balbo seemed overwhelmed by the cheering multitudes. After attending a lengthy ceremony, he attempted to get some sleep by retiring to the privacy of his hotel room. It was to no avail. A herd of reporters followed the Italian Air Minister into his suite and refused to leave.

The seaplanes left Montreal the next morning for Chicago, where thousands of people were waiting for the Italians as the aircraft crossed Lake Michigan and landed adjacent the Century of Progress exhibition grounds.

Massive crowds greeted the adventurers wherever they went in Chicago, and a week later in New York City. Among the numerous official events held in Balbo's honour was a dinner at the White House with President Franklin Roosevelt.

To Balbo, this success seemed to buttress the credibility of Italy's dictatorship. In a telegram to Mussolini he reported that all traces of antipathy towards fascism – which he had experienced in a previous visit to the US – had evaporated. "At the meeting in Madison Square [Garden] a distinct political atmosphere – an atmosphere of passionate Fascism prevails," Balbo wrote.

Balbo's squadron left New York on July 25 on a return flight to Italy. They stopped at Shediac for the night, and headed to Newfoundland the next day. Their destination was Shoal Harbour, a village near Clarenville in Trinity Bay. Balbo was surprised by the sparse, modest amenities awaiting his men in the community. The *Alice* had arrived a week earlier in preparation for the arrival of the air fleet.

Railway carriages had been lined up near the wharf as accommodations for the men, as well as a restaurant car for senior officers. Visitors from several nearby communities, some of whom were camping in tents along the roadside, arrived to greet the Italian planes.

Balbo had expected to leave Newfoundland within five days of arriving. Instead, poor weather kept the Italians waiting for two weeks. With high winds in the forecast for several days, Balbo boarded a train on July 31 and visited St. John's.

In Newfoundland's capital, he and his officers were treated to a sumptuous luncheon hosted by Prime Minister Frederick Alderdice. The prime minister toasted Balbo and led a chorus of "For He's a Jolly Good Fellow" in "a lusty, baritone voice." Balbo also paid a visit to two Italian submarines that were docked in St. John's harbour.

A highlight of the stopover was attending the opening of a new telegraph station at Cabot Tower, a prominent building on Signal Hill overlooking St. John's and the entrance to its harbour. In 1901, from a building adjacent Cabot Tower, Guglielmo Marconi had made a technological breakthrough by receiving the first transatlantic wireless message. Now, more than 30 years later, Balbo was sending a message of greetings from Cabot Tower to the inventor's home in Italy. During his short visit to St. John's, Balbo purchased a pair of sealskin slippers for his infant son, and then returned by train to Shoal Harbour.

Finally, the weather cleared. In the early hours of August 8 the flying boats took off and headed to the Azores. Fog obscured both the sun and the ocean for much of the journey, as it had weeks earlier when they flew to Labrador. One of the seaplanes crashed while landing in the Azores, but nobody was injured in the mishap.

The remaining 23 aircraft continued to Lisbon, Portugal. On August 12, the squadron returned home, landing on the Tiber River in Rome. As Balbo disembarked from his plane, Mussolini was waiting on shore to meet him. Italy's transatlantic heroes marched in a parade through Rome. The king of Italy rewarded Balbo by elevating him to the honourary rank of Air Marshall.

Even though he had become a national hero, Balbo's career as Air Minister came to an abrupt end. Mussolini may have viewed Balbo's popularity as a potential threat to his authority, and in 1934 he appointed the airman as governor of Libya. Here, Balbo coordinated Italy's campaign of modernizing Libyan infrastructure, and oversaw immigration to Italy's "Fourth Shore" by thousands of Italian colonists. His attempts at converting Libyan Moslems to fascism were largely unsuccessful.

Balbo died in June 1940 when his aircraft was shot down over the Libyan port of Tobruk, ironically by anti-aircraft fire from an Italian warship riding at anchor in the harbour. Among the few remaining legacies of his visit to North America is a street in Chicago and a school in Shoal Harbour, both of which bear Balbo's name.

Not all exhibitions of aviation prowess were as successful as Balbo's, as the case of the DO-X flying boat would prove. The limitations of this – the world's first transatlantic passenger aircraft – were not immediately obvious, however. Certainly, those who witnessed the DO-X lumbering across the sky and landing on the ocean surface believed they were witnessing a technological revolution.

This view was shared by executives working for Dornier, the German aircraft manufacturer that launched the first DO-X flying boat in 1929, which hoped that airlines would soon be clamouring to order replicas of the aircraft.

The DO-X was an impressive machine; it was the largest aircraft ever built, with twelve engines mounted on top of a single 48-metre-long wing. Designed by Claude Dornier, the massive seaplane was the epitome of luxury. Its spacious hull held three decks, which contained sleeping quarters, a bar, parlour rooms, a kitchen, and an 18-metre-long dining salon. But the aircraft's size came at a cost. Although it had a large number of engines, the plane remained underpowered; the DO-X weighed 50 tonnes fully loaded and it consumed 1,500 litres of fuel per hour.

Under the command of Captain Friedrich Christiansen the DO-X was supposed to embark upon its maiden crossing of the Atlantic in November 1930. However, engine troubles forced a three-month postponement soon after it left Germany. Then in February 1931, when it arrived in the Canary Islands, the engines had to be overhauled once again. Finally, in June of that year – seven months after leaving Germany – it flew across the South Atlantic and reached the Brazilian island of Fernando Noronha.

Several months later, the DO-X began another flight across the Atlantic, this time flying from New York. On May 20, 1932, with Christiansen at the controls, it headed to the Newfoundland community of Holyrood, but poor weather forced it to land first at the nearby town of Dildo.

The DO-X refuelled in Holyrood and departed Newfoundland on May 22, but the aircraft's chronic unreliability forced Christiansen to land short of Horta in the Azores. It travelled to port under its own power on the ocean surface. Despite the DO-X's problems, enthusiastic crowds greeted the plane as it crossed Europe, stopping in Spain, England, and Germany. By then, however, Dornier had realized that the aircraft's shortcomings made it impractical. Advances in safety and reliability would be necessary before commercial transatlantic flight could become feasible.

Christiansen, who had been a pilot in the First World War, enjoyed a revival of his military career under the Nazi regime, which came to power in 1932. He was promoted to the rank of general and led Germany's military administration in the Netherlands during occupation of that country in the Second World War. He was later charged with war crimes, but was acquitted.

Although Germany failed in its attempts at making flying boats reliable enough to cross the Atlantic Ocean, the country had, since the late 1920s, demonstrated its engineering prowess by building the first airborne ocean liner, the airship.

For a few minutes on October 15, 1928, millions of people in the cities of New York and Washington paused and gazed sky-

The DO-X at Dildo, Trinity Bay, in Newfoundland.

ward. Rush-hour traffic ground to a halt – drivers and passengers seemingly hypnotized. Supreme Court judges stopped their deliberations and dashed to their windows.

The giant *Graf Zeppelin* airship had arrived from Germany. At 236 metres in length, the cigar-shaped dirigible was the largest machine ever made. In Lakehurst, New Jersey, 65,000 spectators greeted the dirigible as it descended gracefully to a landing. Unlike modern lighter-than-air craft that use helium, an inert gas, to provide lift, the *Graf Zeppelin* and its contemporaries were filled with explosive hydrogen gas.

Poor weather and damage to one of the airship's tail fins had slowed its progress across the Atlantic; this resulted in a four-and-a-half-day crossing of the ocean – about two days longer than anticipated. Passengers reported running out of tobacco, but, aside from this annoyance, no significant hardships were endured.

The voyage was billed by the Zeppelin company as a precursor to regular, scheduled commercial flights between Europe, the US, and South America. Not only was the *Graf Zeppelin* important to the business of aviation, the airship would become an important symbol of German superiority in the field of engineering.

Two weeks after arriving in Lakehurst, the *Graf Zeppelin* began its return cruise to Germany. Unlike the westward flight, which followed a route south of the Azores, on its return the airship flew the shorter distance over the treacherous North Atlantic. The dream of airship travel almost ended in disaster over Newfoundland. A massive storm buffeted the *Graf Zeppelin* so much that it was driven backwards towards a group of islands. Hugo Eckner, who headed the Zeppelin firm and was the most experienced airshipman of his era, described the flight as the worst he had experienced.

"This last trip made me think of God," said Eckner, who tried to calm passengers by regularly going to the lounge to feed a pet canary. "I never regarded it as a small thing to cross the ocean in an airship, but never until now have I quite realized what it meant. The storm which caught us over Newfoundland tossed us in a tumult which threatened at times to break the ship in two. Now we know where all those pilots lost their lives. It was over Newfoundland."

Residents of communities in Trinity Bay caught sight of the *Graf Zeppelin* before the airship disappeared over the horizon. It took three days for it to reach the German port of Friedrichshafen.

Following this return flight improvements were made to the airship's engines. In 1929 the *Graf Zeppelin* circumnavigated the globe in a record 21 days. After this achievement regular passenger flights commenced in 1930, with the airship carrying people, freight, and mail across the Atlantic.

The demand for passenger berths convinced Zeppelin to build a new fleet of airships. The convenience of this mode of travel was undeniable; flights from Frankfurt, Germany, and Lakehurst took just 60 hours, a fraction of the time taken by passenger ships. Fares had also become more affordable since the *Graf Zeppelin* began carrying passengers in 1928. In 1936 a new airship, the *Hindenburg*, was launched to meet the burgeoning demand for transatlantic airship travel.

Measuring 245 metres, the length of almost three football fields, the *Hindenburg* was even larger than the *Graf Zepplin*. The design was also innovative; unlike its predecessors, the passenger compartment was partially enclosed within the vessel's hull, an innovation that allowed the airship to carry greater numbers of people. Like the luxurious passenger ships that traversed the world's oceans, the *Hindenburg* boasted lounges, staterooms, and an elegant dining salon. On its maiden voyage, in May 1936, passengers were treated to a piano recital.

The *Hindenburg*, which made ten return trips across the Atlantic in 1936, left a vivid impression with those who witnessed it cruising just 400 metres overhead. Among those who were alarmed by the airship were residents of Colliers, a community on the shore of Conception Bay in Newfoundland: "We looked to the skies and were amazed," said Bridgett Flynn, who had been picking berries when the *Hindenburg* approached. "The Zeppelin was the strangest thing we had ever seen and we thought the world was going to come to an end when it came near."

But by the mid-1930s Germany's airships were no longer just a mode of transportation. With the ascendance of the Nazi Party, under the leadership of Adolf Hitler, the country's fleet of airships became obvious tools for promoting Nazism at home and abroad. Swastikas painted onto the tail fins of both the *Graf Zeppelin* and the *Hindenburg* turned them into gigantic advertisements for Germany's fanatical regime.

In 1936 the two airships were used in a referendum on Nazi leadership; they cruised German cities, dropping party literature. The use of Zeppelin airships for this purpose disgusted Hugo Eckner, president of the company and a well-known opponent of Hitler. When the Nazis took power, German newspapers were initially told to cease using Eckner's name in reports, an order that was quickly rescinded. Eckner had become a highly respected figure in the US, and the German regime was loath to risk a new agreement between Goodyear and Zeppelin that promised to help the airship industry in both countries.

But political pressures – in combination with business decisions by Zeppelin – had already compromised the safety of passengers who were flying in Germany's airships. Beginning in 1927, the US government banned the export of helium in order to guarantee supplies for its own airships. Helium is inherently safer than hydrogen because it is inert and therefore cannot ignite; however, unlike hydrogen, which can be easily produced, helium must be extracted from natural gas. While all natural gas reservoirs contain traces of hydrogen, the US held a virtual worldwide monopoly on helium supplies prior to the Second World War. Although Zeppelin lobbied the US government to gain access to helium, the company had economic reasons for favouring hydrogen. Not only was helium more expensive, it is also a far heavier gas than hydrogen – and therefore a less

The most successful commercial airship, the *Graf Zeppelin*, began its transoceanic career in 1928. Despite having an impeccable safety record, it was immediately retired after the *Hindenburg* exploded in 1937.

Like today's cruise ships, the *Hindenburg* was a luxury vessel, and boasted a piano bar and white tablecloth dining.

effective lift medium.

The aging *Graf Zeppelin* was scheduled for retirement after completing the 1937 season. Between 1928 and 1937 it had flown 1.6 million kilometres without incident, a record of success that greatly reinforced public confidence in airships. From the time it had gone into service in 1928, the *Graf Zeppelin* had become the world's most successful airship; its 590 flights included 143 ocean crossings, mostly between Europe and South America.

Despite its unblemished safety record, the *Graf Zeppelin*'s service life came to a premature end, due to one of the most spectacular accidents of the twentieth century. On May 6, 1937, the *Hindenburg* slowly approached its docking tower at Lakehurst, New Jersey, as it completed its first transatlantic flight of the year. US Navy sailors waited on the ground while mooring ropes were dropped to them from the mammoth airship.

Suddenly, a spark erupted from the stern of the airship. Within seconds, an explosion and sheets of flame ripped into the evening sky. The tanks of hydrogen that had kept the *Hindenburg* aloft during its flight across the Atlantic ignited in rapid succession before a searing onslaught of heat.

The ground crew sprinted to escape the doomed airship as it plunged earthward. Terrified passengers smashed windows and leaped from the gondolas; families waiting to greet them screamed in horror. Within minutes, a showpiece of the Nazi regime had been reduced to a tangle of blackened, twisted metal. Thirty-five people died in the disaster, including the captain, Ernst Lehmann. But these were not the only casualties – the *Hindenburg*'s fate also terminated the era of passenger airships, perhaps forever.

The accident occurred just as the United States and Germany had been planning the construction of several new airships. These investments were being considered despite the loss of several dirigibles to accidents during the previous decade.

Fatal accidents included the *Italia*, an Italian-built craft, which broke apart over Norway during a scientific expedition in 1928, killing several men. British airships, including the *R-100*, which attracted hundreds of thousands of visitors in Quebec, southern Ontario, and New York state following a transatlantic voyage in July-August 1930, were permanently grounded two months later when the *R-101* crashed during its maiden voyage over France, bound for Karachi, killing all but six of its 54 passengers and crew.

The US had also experienced disasters involving Navy airships,

The *Hindenburg* exploded without warning while approaching a docking tower at Lakehurst, New Jersey, on May 6, 1937. US Navy sailors, preparing to grab tether lines on the ground, ran for safety as the airship descended in flames.

starting with the *Shenandoah*, which broke apart over Ohio in 1925. Then, in 1933, only three of 76 people survived when the *Akron* plunged into the ocean near Atlantic City, New Jersey. Similarly, two years later high winds tore the *Macon* to shreds, killing two people. In contrast, Germany had not recorded any serious accidents involving passenger airships built and operated by the Zeppelin firm since the end of the First World War.

Germany did not immediately terminate its airship program following the *Hindenburg* accident. The government intensified its efforts at obtaining helium in order to provide lift for another airship that had been set to enter the transatlantic service. But with a war looming in Europe, the US government had become suspicious about German motives and the moratorium on helium exports remained in place.

In June 1937 the *Graf Zeppelin* was deflated and placed in a hangar one month after the spectacular destruction of the *Hindenburg*. With the outbreak of war in 1939, the skeleton of the *Graf Zeppelin* was dismantled and melted to manufacture weapons.

Passenger airships would no longer cruise the world's skies. They were replaced after the Second World War by a new generation of airliners – which proved to be faster, safer, and more economical.

CHAPTER 7

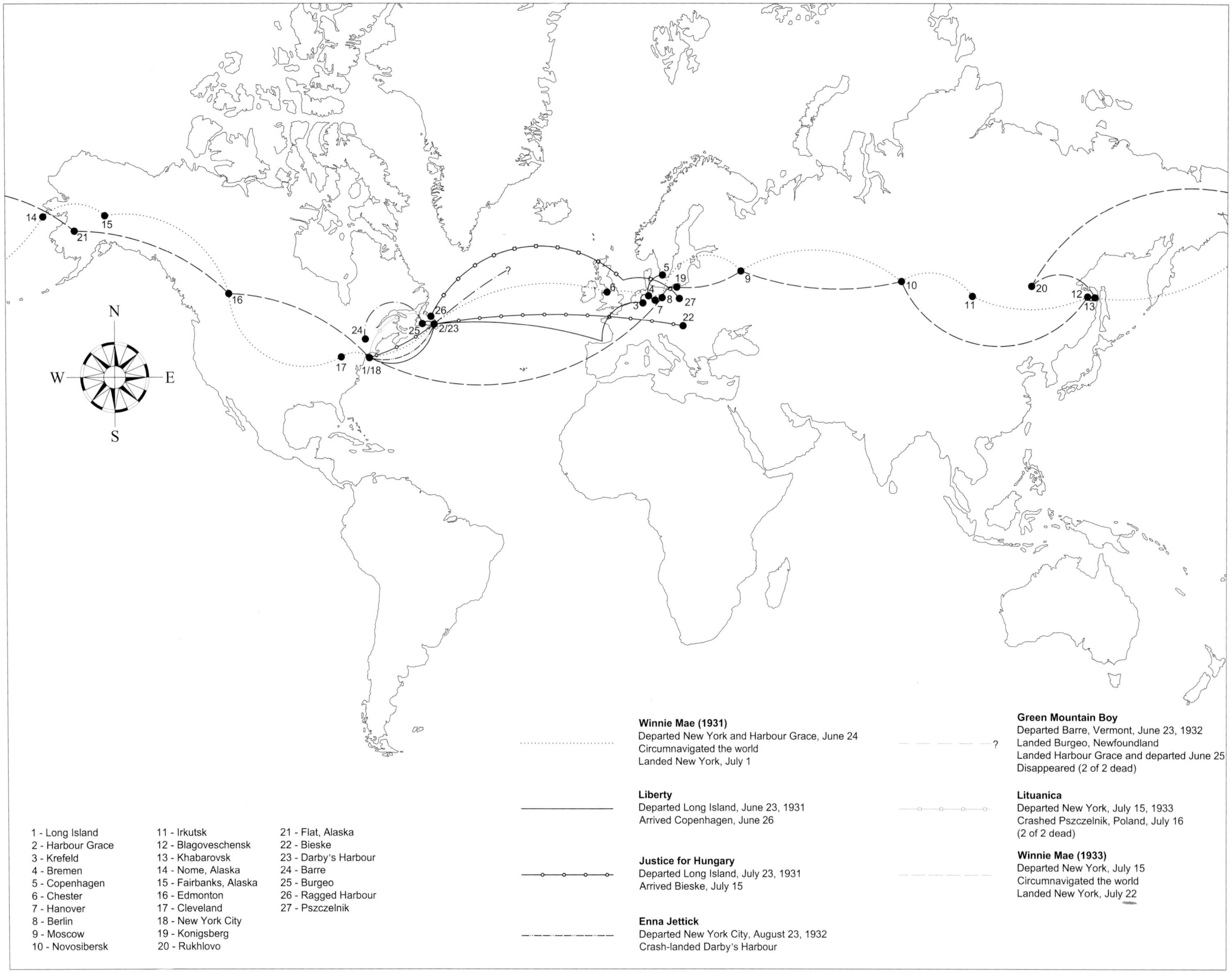
N
S
E
W
1 - Long Island
2 - Harbour Grace
3 - Krefeld
4 - Bremen
5 - Copenhagen
6 - Chester
7 - Hanover
8 - Berlin
9 - Moscow
10 - Novosibersk
11 - Irkutsk
12 - Blagoveschensk
13 - Khabarovsk
14 - Nome, Alaska
15 - Fairbanks, Alaska
16 - Edmonton
17 - Cleveland
18 - New York City
19 - Konigsberg
20 - Rukhlovo
21 - Flat, Alaska
22 - Bieske
23 - Darby's Harbour
24 - Barre
25 - Burgeo
26 - Ragged Harbour
27 - Pszczelnik
Winnie Mae (1931)
Departed New York and Harbour Grace, June 24
Circumnavigated the world
Landed New York, July 1
Liberty
Departed Long Island, June 23, 1931
Arrived Copenhagen, June 26
Justice for Hungary
Departed Long Island, July 23, 1931
Arrived Bieske, July 15
Enna Jettick
Departed New York City, August 23, 1932
Crash-landed Darby's Harbour
Green Mountain Boy
Departed Barre, Vermont, June 23, 1932
Landed Burgeo, Newfoundland
Landed Harbour Grace and departed June 25
Disappeared (2 of 2 dead)
Lituanica
Departed New York, July 15, 1933
Crashed Pszczelnik, Poland, July 16
(2 of 2 dead)
Winnie Mae (1933)
Departed New York, July 15
Circumnavigated the world
Landed New York, July 22

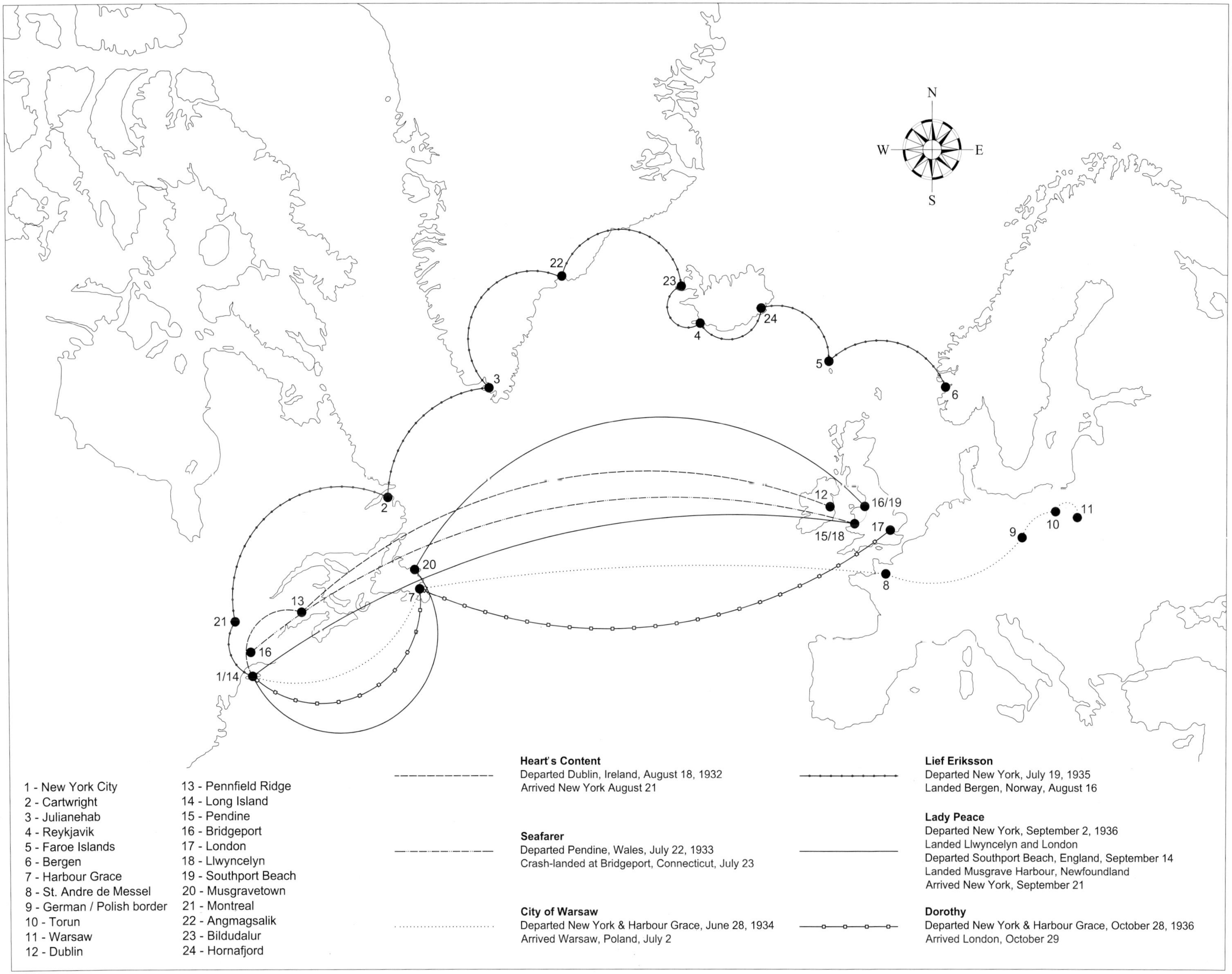
N
W
E
S
1 - New York City
2 - Cartwright
3 - Julianehab
4 - Reykjavik
5 - Faroe Islands
6 - Bergen
7 - Harbour Grace
8 - St. Andre de Messel
9 - German / Polish border
10 - Torun
11 - Warsaw
12 - Dublin
13 - Pennfield Ridge
14 - Long Island
15 - Pendine
16 - Bridgeport
17 - London
18 - Llwyncelyn
19 - Southport Beach
20 - Musgravetown
21 - Montreal
22 - Angmagsalik
23 - Bildudalur
24 - Hornafjord
Heart's Content
Departed Dublin, Ireland, August 18, 1932
Arrived New York August 21
Seafarer
Departed Pendine, Wales, July 22, 1933
Crash-landed at Bridgeport, Connecticut, July 23
City of Warsaw
Departed New York & Harbour Grace, June 28, 1934
Arrived Warsaw, Poland, July 2
Lief Eriksson
Departed New York, July 19, 1935
Landed Bergen, Norway, August 16
Lady Peace
Departed New York, September 2, 1936
Landed Llwyncelyn and London
Departed Southport Beach, England, September 14
Landed Musgrave Harbour, Newfoundland
Arrived New York, September 21
Dorothy
Departed New York & Harbour Grace, October 28, 1936
Arrived London, October 29

[illegible] set of the Great [illegible] on Harbour Grace [illegible] ome the main depar- [illegible] val point for transat- [illegible] ft. So important was [illegible] nity to aviators that in June [illegible] two Europe-bound aircraft took off from Harbour Grace on the same day – an unprecedented event for the community during the early days of long-distance aviation.

A Bellanca monoplane, named *Liberty*, departed Floyd Bennett Field in Brooklyn on June 22, and arrived in Harbour Grace later that day. Its two crewmen were hoping to complete the first non-stop flight to Copenhagen, Denmark, from North America.

The pilot, Holger Hoiriis, was a native of Denmark who had immigrated to the US following the First World War. Accompanying Hoiriis was a passenger, the plane's owner, Otto Hillig, a German immigrant who became a wealthy portrait photographer in New York State. Hillig had named the plane after Liberty, New York, the city that became his home.

According to newspaper reports, Hillig had purchased his plane using the proceeds from an out-of-court settlement reached with operators of the *Graf Zeppelin* airship in 1928. He had filed a lawsuit against the company when his reservation on a transatlantic flight was cancelled without warning.

On June 24, almost two days after the *Liberty* had landed at Harbour Grace, it was joined by a second plane, the *Winnie Mae*, which had left Roosevelt Field, Long Island, several hours earlier. But pilots Wiley Post and Howard Gatty had even a grander ambition than Hoiriis and Hillig: they were planning to fly around the world, and to break the record of 21 days that had been established in 1929 by the *Graf Zeppelin*.

This round-the-world flight would establish the then little-known Post as one of aviation's greatest pioneers. But Post, born part-Cherokee Indian, was perhaps an unlikely candidate to achieve greatness. In his youth Post had served time in jail for stealing a car, and later found work in the Oklahoma oil fields, where he lost an eye in an industrial accident. But this mishap had one benefit: Post was able to use the proceeds from a financial settlement from his employer to buy an aircraft and pursue his love for flying.

In the mid-1920s, Post became the personal pilot of Oklahoma oil tycoon, F.C. Hall. In 1930, Hall bought a Lockheed Vega and named it *Winnie Mae*. The Vega had become popular with many pilots, among them Amelia Earhart, who made several record-setting flights with her machine. Hall agreed to finance Post's dream of flying around the world and hired Gatty, an Australian, as co-pilot.

Although the *Winnie May* had started its journey after the *Liberty*, Post and Garry were determined to beat their rival in crossing the Atlantic. They quickly refuelled their plane and took off from Harbour Grace first just four hours after arriving in the community. The Atlantic proved to be the most dangerous part of the world flight for Post and Gatty. Soon after taking off from Harbour Grace, the *Winnie Mae* was battling the same weather conditions that had plagued many transatlantic aviators.

"There is one thing I don't want to do right away again, and that is to fly the Atlantic," Post recalled. "For three hours I could not see the engine at all and I just had to keep my eyes glued to the instruments. And it was only by the grace of God that we found that hole over Bangor in Wales, came down and found out where we were."

The *Winnie Mae* then stopped in England and Germany, before flying for several days over the vast Siberian wilderness. Post said the highlight of his world journey occurred at Irkustk in central Siberia, where most of the city's inhabitants converged to meet the plane. The only English speaker was a sixteen-year-old girl, who did her best to translate.

"The poor girl had spoken no English for six years ... and mixed a strong cockney accent with Russian at first, but gradually we got to understanding each other better and we talked about the various countries we had visited and had a good chat," Post wrote.

On June 30 they reached Fairbanks, Alaska, and on the same day flew 2,330 kilometres to Edmonton, Alberta. There, the *Winnie Mae* was greeted by 2,000 people, many of whom broke through a police line to greet the flyers. The plane had landed at Blatchford Field, the first publicly licensed airfield in Canada – later to be renamed Edmonton City Centre Airport. The rain had turned the airstrip into mud, so the next morning the *Winnie Mae* was towed to the city's main road, Portage Avenue – now Kingsway Avenue – where it took off and headed east.

On July 1, Post and Gatty arrived where they had started, at Roosevelt Field. They had flown around the world in eight days and sixteen hours, smashing the *Graf Zeppelin*'s time of 21 days and five hours.

Although the *Liberty* had lost the transatlantic race to the *Winnie Mae*, Hoiriis and Hillig were determined to complete their non-stop flight to Copenhagen. They departed Harbour Grace before sunrise on June 24 and soon met one of the greatest hazards to aviation – fog. However, unlike many aviators who had attempted to fly across the Atlantic, Hoiriis had prepared for this particular challenge: in the weeks prior to the flight he had trained himself for flying in fog by covering the windows of the *Liberty* with paper and flying by instrumentation alone. Such precautions, often omitted by other flyers prior to attempting transatlantic crossings, almost certainly saved his life and Hillig's.

Although Hoiriis's training provided him with the skills necessary to keep the plane from crashing into the ocean and veering wildly off course, he was unable to determine with precision the plane's location. For fifteen hours the *Liberty* flew blind and against strong winds. Eventually, the skies opened. "We finally discovered a hole in the clouds which had kept us for what seemed an eternity completely secluded from the world," Hillig recounted. "The changed situation, however, brought no relaxation, since we had not the remotest idea where we were and hardly any way to find out."

The men flew along the coast of northern Europe for several hours, until they finally found an airport, at Krefeld in Germany's Rhineland district. They were too tired to continue further. They landed and waited until the next day to fly the remaining 320 kilometres to Denmark. As they arrived in Copenhagen Hoiriis and Hillig were mobbed by 50,000 cheering spectators.

Two years after the *Winnie Mae* set the round-the-world mark, Post decided to lower his record again – but this time he would to do it flying solo. He took off from Floyd Bennett Field, New York, in July 1933 to begin his quest. This time Post had taken advantage of technical advancements by installing an autopilot and radio compass inside the aircraft. He saved time by flying directly from New York to Berlin, but subsequently lost many hours due to equipment malfunctions, minor accidents, and losing his way over Alaska. On July 22 he once again refuelled the *Winnie Mae* in Edmonton and continued on to New York. Despite the delays, Post had circled the globe in a record time of seven days and nineteen hours.

Post then focused on high-altitude flight, and, in partnership with B.F. Goodrich, he designed a pressurized aviator suit. Post wore one when he took the *Winnie Mae* to an altitude of 15.2 kilometres, reaching the sub-stratosphere. In doing so, he discovered the existence of the jet stream.

His career was assisted by a friendship with American humorist Will Rogers, whom Post met in

Harbour Grace hosted two teams of transatlantic flyers concurrently in June 1930. The *Winnie Mae* (left) succeeded in flying around the world, while the *Liberty* (right) flew non-stop to Copenhagen, Denmark.

1925 while flying the writer to a rodeo. Rogers subsequently gave the aviator considerable free publicity, and in return Post provided him with flights across North America. It was on one of these excursions that they ended their lives; in August 1935, as Post and Rogers took off from Point Barrow, Alaska, the engine of their plane stalled – it careened back to earth, killing both men. Rogers, whose life was marked by his success as an actor, author, pundit, and philosopher, was memorialized by the *New York Times* as "the most widely-known citizen of the US … and the best beloved."

Post was motivated by several factors, including money, prestige, and a competitive drive, to break speed and endurance records. Patriotic fervour inspired many other attempts at flying across the Atlantic, but none would match the absurdity of a Hungarian effort in 1931 that almost ended in a shoot-out.

The goal of flying from Harbour Grace to Budapest was designed to bring attention to the dismantlement of Hungary through the Treaty of Trianon after the First World War. Many Hungarians remained

Above: A radio reporter (kneeling) records a meeting of pioneer airmen in Harbour Grace, as Holger Hoiriis (centre right) shakes Howard Gatty's hand. Between them is Post, with Otto Hillig (in an aviator hat).

Below left: Humorist and actor Will Rogers, starring in the 1929 film, *They Had to See Paris*. Below: Upon their arrival back in the United States, Gatty and Post were accorded a tickertape parade along the heart of New York's financial district in July 1931.

bitter about losing two-thirds of their country to the newly established nations of Czechoslovakia, Romania, and Yugoslavia.

Among them were two Hungarian reserve pilots, Sándor Wilczek and György Endres, who would bring attention to this issue by flying non-stop from Newfoundland to Hungary. Both men had recently emigrated to North America, but like many Hungarian expatriates they identified closely with their native country. In fact, when Wilczek immigrated after the war, he changed his surname to Magyar, which literally means a native of Hungary. Endres, who served with the Austro-Hungarian air force during the war, had been Magyar's flight instructor.

Magyar moved to Windsor, Ontario, where he boarded with a clergyman, William Molnar, who encouraged him to follow his dream of flying to his homeland. He claimed to have been convinced when Molnar opened the Bible at random and came upon the line, "These things ye shall do unto the Lord ..."

A Michigan meat-packing businessman, Emil Szalay, raised US $20,000 by mortgaging his business; he turned the proceeds over to Endres, who purchased a Lockheed Sirius aircraft, identical to one owned by Charles and Anne Lindbergh. Szalay said the commitment fulfilled a promise made to his father many years earlier that he would "do something for Hungary."

A transatlantic flight with blatant political overtones was that of the *Justice for Hungary*. Its financiers used the flight to draw attention to Hungary's territorial losses following the First World War. The flyers were greeted by large crowds when they reached Hungary.

Aside from bringing attention to their cause, Endres and Magyar were also hoping to win the prize money offered by one of Hungary's most influential European supporters, Harold Harmsworth, who was given the title Viscount Rothermere in 1919. In 1917, Prime Minister Lloyd George had appointed the British newspaper tycoon as President of the Air Council, which was the precursor to the Royal Air Force.

Harmsworth and his brother Alfred owned the *Daily Mail* and several other influential newspapers; by 1926 Rothermere had become one of the richest men in Great Britain. During the 1930s he became a prominent supporter of appeasement with Adolf Hitler.

The Harmsworths were well known in Newfoundland where they had operated a pulp and paper mill in Grand Falls since 1909; their Anglo-American Development Company provided a cheap supply of newsprint for their newspapers. Although the mill would later be sold, the family's association with Newfoundland was continued by Esmond Harmsworth, Harold's son, who became chancellor of Memorial University of Newfoundland in 1952.

From the early days of aviation the Harmsworths had been synonymous with aircraft, offering cash prizes for several pioneering feats. Among the most notable was a £10,000 *Daily Mail* award for flying non-stop across the Atlantic, which was won by Alcock and

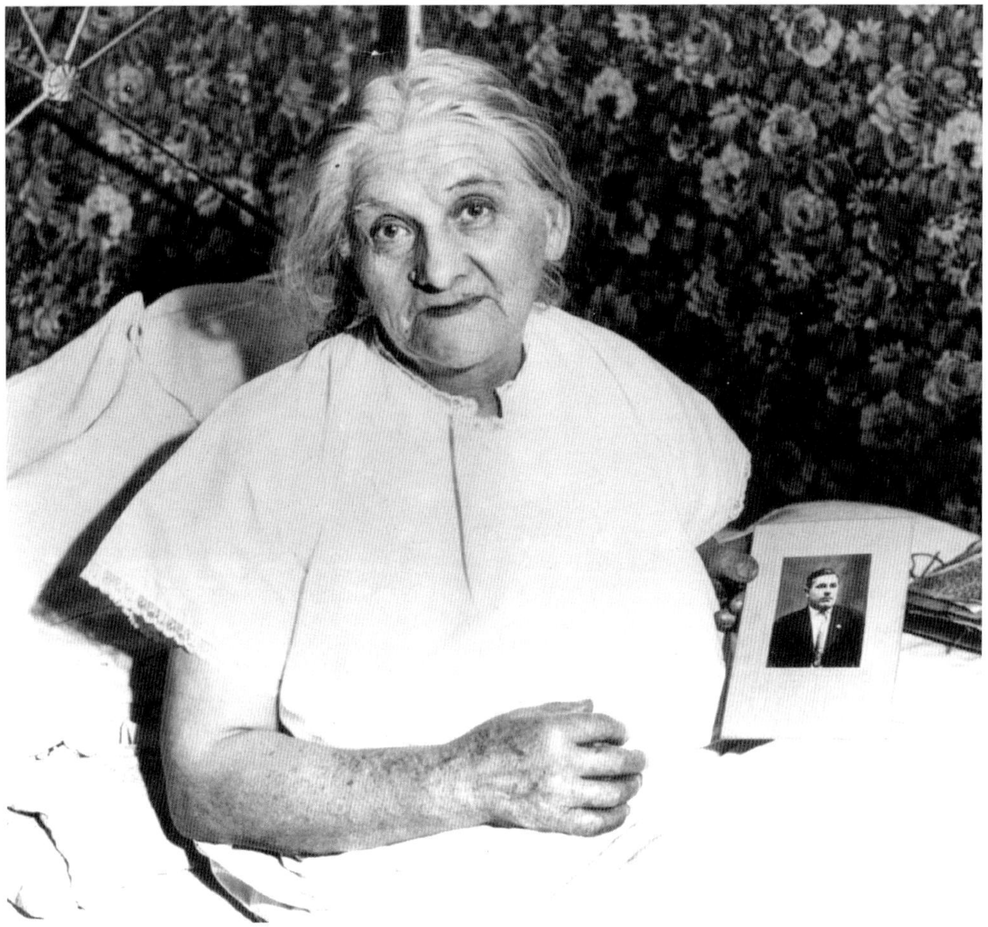

Family members of transoceanic flyers endured the stress of waiting for the safe arrival of their loved ones. Elizabeth Endres is pictured in her home on July 15, 1931, anxiously awaiting news about her son, György Endres, who, with Sándor Wilczek, was flying over the Atlantic, en route to Budapest.

Brown in 1919.

After the war Rothermere became a supporter of the Hungarian cause, and in 1930 he offered US $10,000 to the first Hungarian to fly non-stop from North America to Budapest, Hungary's capital city. Rothermere took a particular interest in the bid by Endres and Magyar, even choosing a name for their plane, *Justice for Hungary*.

The aircraft, with Endres and Magyar in the cockpit, left Roosevelt Field on July 13 and arrived in Harbour Grace later that evening. They were greeted by a small group of Hungarians who had travelled to Newfoundland to support their compatriots. Several prominent members of the St. John's political and business elite joined them in Harbour Grace. Two days later the *Justice for Hungary* took off for Europe. Although the transatlantic leg of the flight was uneventful, the plane had used more fuel than anticipated. A little over 26 hours after leaving Harbour Grace Endres and Magyar found themselves with empty tanks only 32 kilometres from Budapest, and they had no choice but to crash-land the plane in a field near the town of Bieske. Rothermere awarded them the prize money, even though the *Justice for Hungary* had come short of reaching Budapest.

Hungarians initially rejoiced at the achievement. The nation's armed forces awarded promotions to both Endres and Magyar, and large crowds greeted the pilots at public events. But public feuding soon replaced the initial goodwill, as questions arose about Magyar's past. It was suggested that their hero had changed his name and fled Hungary in order to avoid repercussions arising from "misconduct." The nature of this claim is unclear. With his honour besmirched, Magyar responded by challenging six members of Budapest's sporting society to a duel. Whether he intended to fight them in succession or as a group is not known, but a settlement was reached without resorting to violence.

Trouble arose once again, however, when Szalay sold the *Justice for Hungary*. After reimbursing himself for expenses associated with the flight, he proposed splitting the surplus between the two flyers. By now, however, Endres was no longer on speaking terms with Magyar, and told Szalay that he alone was entitled to the money. This enraged Magyar, who challenged Endres to yet another duel. Endres accepted. According to the two men, fighting a duel was the only acceptable way for Hungarian military officers to settle such disputes.

At this point the Hungarian Aviation Society intervened and pleaded with the men to call off the duel. The aviators eventually conceded and Magyar returned to the United States. But Endres would not enjoy his success for long; he was killed in May 1932 while landing an aircraft in Rome. He had travelled there to attend an aviators' convention, hosted by the Italian president, Benito Mussolini.

By 1932 the Depression had reached its lowest point, as North America and Europe buckled under the weight of mass unemployment and economic devastation. Perhaps in response to these economic difficulties the public yearned for entertainment in all its forms, including stunt flying, air races, and barnstorming. But the cash prizes that fuelled pioneer flight throughout the 1920s had become increasingly scarce due to difficulties obtaining sponsors from cash-starved businesses. In addition, there were few "firsts" left to be achieved in conventional aviation.

So there was considerable interest within the aviation fraternity when a Norwegian newspaper offered US $10,000 for the first non-stop flight to Oslo from North America. In August 1932 a race began between two teams of flyers.

One team consisted of Carl

Clyde Lee was a daredevil who thrilled crowds at air shows with his "wingwalker" stunt. His decision to fly from Harbour Grace (bottom) to Norway would prove to be his undoing, however, as both he and co-pilot John Bochkon disappeared over the north Atlantic. Left: Bochkon and Lee.

"It's the money I need. A good new engine, and it would be a cinch."
— Clyde Lee, on flying across the Atlantic

Petersen and Thor Solberg, both of whom had been born in Norway and had moved to the United States. Petersen previously travelled twice to the Antarctic with Richard Byrd and played a significant role in advancing the science of radio communications. In 1932 Petersen, who owned a Bellanca aircraft, convinced the Enna Jettick shoe company to sponsor his flight. He named the plane *Enna Jettick*.

Their competition was a Wisconsin pilot, Clyde Lee, and his Norwegian-born navigator, John Bochkon. Lee was known as "Daredevil Lee" because he performed stunts such as walking along the wing of an aircraft while it was in flight. Lee was unable to obtain financial support for the transatlantic journey in Wisconsin. But a Vermont cement manufacturer agreed to pay the expenses, provided the Stinson monoplane was christened *Green Mountain Boy*. This name was chosen to honour a group of soldiers from Vermont who had fought in the American Revolutionary War.

On August 23 the transatlantic race began, with the *Enna Jettick* starting from Floyd Bennett Field, New York, and the *Green Mountain Boy* from Barre, Vermont. A further dramatic touch was added when both crews disclosed that they were planning a stop at Harbour Grace before heading to Norway.

But the race ended only a few hours after it began. Navigation became difficult, as the hazards of rain, fog, and snow enveloped both aircraft as they flew over southern Newfoundland. Lee managed to land the *Green Mountain Boy* on a beach in the community of Burgeo. The crew of the *Enna Jettick* were not so lucky. As the sky darkened towards evening, Petersen and Solberg looked for a safe open space to land their plane.

They spotted lights over Placentia Bay, but, while descending, Solberg suddenly noticed a hillside looming into view. He banked sharply to avoid disaster, but in doing so one of the plane's wings struck the water. The *Enna Jettick* finished its journey tail-up in the fishing village of Darby's Harbour. Petersen and Solberg survived the accident unscathed.

The next day, August 24, the *Green Mountain Boy* left Burgeo and landed at Harbour Grace. A day later, Lee and Bochkon set out for Norway. Although weather forecasts provided to the pilots were favourable, in reality a massive fog bank had enveloped much of northern Europe. The fog was so dense that all flights in Great Britain had been cancelled. In Wales, there were reports of people hearing an aircraft engine, and it was speculated that it belonged to the *Green Mountain Boy* since all flights had been grounded due to the fog. There was no confirmation, however, and neither the plane nor Lee and Bochkon were seen again.

But Solberg would get a second chance to reach Norway by air. In

1935 he and Paul Oscanyan successfully flew a Loening amphibious aircraft, named *Leif Eiriksson*, on a northern route to Bergen. The 8,500-kilometre journey from New York included stops in the Labrador community of Cartwright, Greenland, Iceland, and the Faroe Islands.

In the mid-1930s several other European nations became the focus of transatlantic aviation. On July 15, 1933, Lithuanian American pilots Stephan Darius and Stanley Girenas took off from Floyd Bennett Field in the *Lituanica*, bound for Kaunas, Lithuania. For several hours, the flight proceeded without incident, as the *Lituanica* passed over Newfoundland, but bad weather forced them north of their planned flight path over the Atlantic. Although the pilots managed to get back on course over Germany, they were killed when their plane crashed near the town of Pszczelnik, Poland – then known as Soldin, Germany. Since no mechanical failures were found and the plane still had fuel on board, investigators theorized that the men had fallen asleep.

The race to fly non-stop to Poland had begun inauspiciously a year earlier, in 1932, when Stanislaus Hausner was forced to ditch his plane in the ocean and wait several days inside the wreck before being rescued. In 1934 it seemed that Polish honour would be restored by two native sons – that is, until US authorities decided to probe the business activities of the erstwhile heroes.

Crash Ends Trans-Atlantic Race

A RACE BETWEEN TWO MONOPLANES to cross the Atlantic came to an end with the wreck of one ship at Placentia Bay, Newfoundland. The second plane landed far west of Harbor Grace, where both had planned to refuel, but missed their objective when storms and fog obscured landing beacons. Pictures show John Bochkon and Clyde A. Lee, crew of the "Green Mountain Boy," which landed safely and which is continuing its flight from Barre, Vermont, to Oslo, Norway. INSET, Thor Solberg and Carl Petersen, crew of the plane which crashed.

Benjamin and Joseph Adamowicz were teenagers when their widowed mother moved with them to Brooklyn, New York, from Poland in 1911. The boys, who found work in a bottling plant, later opened their own soda water bottling business on Long Island.

Like many people who became enamoured with aviation, the Adamowicz brothers were inspired by the New York-Paris race in 1927. They were among thousands of people who visited airfields across the United States and paid money to experience the thrill of flying. After being a passenger on one of these short flights, Benjamin reportedly told Joseph, "I can do that." Both men then earned their pilot's license.

Above left: An image from the *Toronto Telegram* of the men competing to reach Oslo, Norway, (left to right) John Bochkon and Clyde Lee (insert), Thor Solberg, and Carl Petersen. Above: The *Enna Jettick* ended its journey at Darby's Harbour, Newfoundland. The flight was sponsored by the Enna Jettick shoe company, which also promoted its wares using a trailer designed by aircraft engineer Glenn Curtiss – builder of the Navy Curtiss flying boats that reached the Azores in 1919.

While many people dreamed of flying across the Atlantic, the Adamowicz brothers were among the few to actively pursue this quest. In 1933, they hired Holger Hoiriis, who had flown to Denmark from Newfoundland in the *Liberty* two years earlier. Hoiriis gave them further instruction, including valuable advice on flying the treacherous skies of the North Atlantic.

The Adamowicz brothers purchased the *Liberty* from Hoiriis and renamed it *White Eagle* – a white eagle being on Poland's official coat of arms. On August 7, 1933, they and pilot Emile Bergen flew to Harbour Grace. The following day the three men boarded the *White Eagle* and prepared for takeoff. However, as the plane gained speed it suddenly veered off the runway and crashed into the brush. The three men had not been seriously injured, but the plane was badly damaged and they were forced to return to New York.

Despite this failure, the brothers were determined to try again the following year. After the plane had been repaired, Hoiriis taught them how to fly using instruments alone – a lifesaving skill when travelling through fog and darkness.

The Adamowicz brothers damaged the *White Eagle* on takeoff from Harbour Grace in 1933. Their Bellanca aircraft had previously flown to Denmark in 1931 under the name *Liberty*.

On June 28, 1934, Hoiriis accompanied Benjamin and Joseph to Harbour Grace. The plane, which had been renamed *City of Warsaw*, took off without incident the next day with only the brothers on board. Hoiriis returned to New York.

Unlike many of their predecessors, the brothers experienced only minor difficulties crossing the Grand Banks of Newfoundland. Their luck ran out, however, as the *City of Warsaw* neared France, when a severe storm drove them off course. After flying blindly through fog for three hours, they finally found land. The plane was down to its last few minutes of fuel when the pilots sighted what they thought was Ireland. They were wrong. The plane had actually landed in a pasture at the village of St. André de Messel, 80 kilometres from Caen in northern France.

The brothers flew to Paris the next day and hoped to reach Warsaw after a few hours of flight. News of their imminent arrival swept through Poland, and some 15,000 people waited anxiously for the Adamowiczs' arrival at Warsaw's military airport. Difficulties continued to plague the aviators, however, as the plane ran out of fuel over Germany, forcing them to land in a potato field near the Polish border.

Finally, on July 2 the plane landed in Warsaw, although not before it made another unplanned stop in the Polish city of Torun. A squadron of bombers accompanied the *City of Warsaw* as it approached the capital. The US ambassador greeted them at the airport, after which the brothers were taken on a procession to City Hall.

The Adamowicz brothers enjoyed two months basking as celebrities in the nation of their birth. They even profited from their exploits by selling the *City of Warsaw* to a museum.

On September 23 the brothers boarded a passenger ship at Gdynia, and bade farewell to thousands of cheering spectators. "We're going back to the soda water business in Brooklyn," Joseph said before they left Poland. "But we won't neglect flying, and we hope to fly back to Warsaw without stopping en-route."

But their triumphant departure masked trouble at home. A day before the Adamowicz brothers left Poland, New York police had swooped into their bottling plant, which was being run during their absence by another brother, Bronislaw. It seems that soda water was not the only beverage being bottled on the premises; the authorities announced that they had raided the largest illegal still ever uncovered on Long Island.

The three brothers protested their innocence in the face of overwhelming evidence. They claimed that a man named Schwartz had rented part of their business; however, investigators could find no record of this person. A jury was unable to reach a verdict in the first trial, but in April 1935 all three were convicted. They were sentenced to fifteen months in jail.

The motivations of those such as the Adamowicz brothers who attempted to fly across the Atlantic during the early years of flight fell loosely into the following categories: money, patriotism, personal glory, and thirst for adventure.

In 1936, Scottish flying hero James Mollison would add another reason – revenge. Mollison had burst into prominence in 1931 by

After repairing their plane, which was re-christened the *City of Warsaw*, the Adamowicz brothers returned to Harbour Grace in 1934. Below: Benjamin and Joseph Adamowicz (fourth and fifth from left) reached Europe on their second try, but had to stop in France after running low on fuel.

setting a record of almost nine days for flying to England from Australia. Rather fittingly, he landed near Hastings, where William the Conquerer first set foot in 1066 – the last successful military invasion of England.

But it was Mollison's love life that captured the public's imagination. The consummate ladies' man had been engaged to a young aristocrat until he met Amy Johnson, a former secretary who shared his love for flying. Johnson had completed pilot training in the late 1920s, and soon began setting records of her own. In 1930, she became the first woman to fly solo from Great Britain to Australia, and in 1932 she broke the speed record for flying to South Africa from Britain.

She and Mollison were soon married. To celebrate their nuptials, Mollison announced that he would become the first person to achieve a solo flight westward across the Atlantic, from Ireland to New York City. The newlywed aptly named his aircraft *Heart's Content*. Mollison came close to succeeding in his quest, but his plane ran out of fuel before reaching New York. He had flown from Ireland to Pennfield Ridge, New Brunswick, a community near Saint John, in 30 hours.

In 1933 the record-setting couple embarked upon their most ambitious project, flying the Atlantic together, from Wales to New York. The duo almost made it; their aircraft, named *Seafarer*, ran out of fuel at Bridgeport, Connecticut, only minutes from New York City. Both pilots suffered minor injuries as the plane plowed into the ground and overturned.

But the public image of a happy couple, participating in wild adventures, was a facade. Their marriage was foundering, as Mollison enjoyed the company of women other than his wife. The first public hint of discord came in May 1936 when Johnson flew solo on a return flight between England and Cape Town. In doing so she demolished the previous record set by her husband by making the journey in just

James Mollison and the *Dorothy*, a Bellanca Flash, in Harbour Grace prior to setting a speed record for flying across the Atlantic Ocean.

four days and sixteen hours. Mollison was not among the large crowd that greeted Johnson upon her return.

After surviving an air crash in October that year Johnson publicly disclosed that she and her husband had separated. This unwelcome publicity infuriated Mollison. But the scorned husband responded in a unique fashion: he would attempt to set a new speed record for flying across the Atlantic, and then immediately smash the mark set months earlier by his estranged wife for a return flight between England and Cape Town.

On October 28, just days after Johnson's revelation about their marital difficulties, Mollison flew from New York and arrived in Harbour Grace hours later. He was flying a newly acquired Bellanca 28-70 monoplane which he named *Dorothy* in honour of Dorothy Ward, an actress with whom he was having a relationship. The flamboyant aviator climbed from the cockpit in Harbour Grace wearing a stylish dinner jacket, but he was in no mood to discuss his flight from New York. "I am cold and tired," was all he told reporters before seeking refuge in his hotel room.

Mollison took off the next day. He arrived safely at Croydon airport thirteen hours and seventeen minutes later, thereby setting a new speed record for crossing the Atlantic. In doing so, he had also become the first person to fly across the ocean four times. True to form, Mollison displayed his penchant for high living after stepping from his plane. "I am so damned tired," he said, while caressing a lucky rabbit's foot, given to him by a girl in Harbour Grace. "I am going to have a large scotch and soda." As drinking establishments would not open for several hours, Mollison had to settle for ginger beer.

He may have crossed the Atlantic in record time, but the grudge flight to South Africa and back would have to wait. First, Mollison decided to visit Ward at her home in Edinburgh. Then, the Bellanca Aircraft Corporation seized his plane, pending payment of the balance outstanding on purchase of the machine. Finally, a month after arriving in London, Mollison had settled with Bellanca and was ready to challenge his wife's record. Unlike Johnson, who flew solo, he hired a co-pilot, Edouard Cornig-lion-Molinier. He and Mollison had previously flown together while conducting aerial surveys for archeologists in northern Africa. The French flying ace led a remarkable military career, winning the Croix de Guerre in both world wars.

The two men took off from Croydon Airport on November 29. Despite some delays, they remained on a record-setting pace – until losing their way in South Africa. They were listed as missing for several days, until being found just 160 kilometres from Cape Town. Johnson's record had withstood the challenge.

With the advent of the Second World War both Mollison and Johnson joined the British Air Transport Authority, ferrying aircraft from factories to air bases.

Amy Johnson exacted revenge against her cheating husband, James Mollison, by beating his record for a return flight between England and South Africa.

Next page: The *Dorothy* flying over Brooklyn, New York, prior to Mollison's 1936 flight.

While flying a mission in January 1941, Johnson was forced to ditch into the Thames estuary after overshooting her destination and running out of fuel. She drowned before a rescue craft could reach her.

Mollison survived the war, but alcoholism exacted its toll on his health. He died in 1959 at the age of 54. Corniglion-Molinier would continue to lead a successful, varied career; he became a film producer after the war, and then entered politics, where he served in several cabinet posts.

By the mid-1930s, advances in aviation technology and aircraft safety had greatly reduced the element of danger for pilots and passengers. Airlines throughout North America and Europe were competing for airmail contracts and several companies offered passenger flights between major cities to the few who could afford tickets.

However, flying across the world's oceans remained exceptionally dangerous, as one of America's leading entertainers would find out in 1936. Nightclub singer Harry Richman had soared to fame during the 1920s, sharing the stage with such stars as Mae West. The "King of Broadway" enjoyed a string of popular records, as well as a torrid relationship with the original "It Girl," Clara Bow.

Richman's success gave him the means to engage in the expensive hobby of aviation. He became friends with American First World War flying ace, Eddie Rickenbacker, who had since become a senior executive with Eastern Air Lines. The two men had more than aviation in common; during the First World War both had changed the spelling of their Germanic surnames – Reichman and Reichenbacker.

Richman was determined to achieve one of the few unclaimed transatlantic records: the first round-trip flight by plane between the US and Europe. On September 2, 1936, he and Eastern Air Lines pilot Henry T. (Dick) Merrill took off from Floyd Bennett Field in Brooklyn, New York, bound for London. Richman stated that he would have tea in the British capital, and then dine on Broadway after returning to New York the following evening.

For the voyage, Richman and Merrill flew a Vultee VI-A, a single-engine aircraft used by American Air Lines for its US passenger routes in the early 1930s. Richman drew attention to rising political tensions in Europe by naming his plane *Lady Peace*. He also underlined the danger of the flight by stowing a load of 41,000 ping-pong balls into cavities within the wings of the plane, to make the craft more buoyant in case he and Merrill were forced to ditch it at sea.

Although the flight to Europe was largely uneventful, Merrill lost his way as the *Lady Peace* neared land. The plane had run low on fuel by the time the men finally sighted the British Isles, following an eighteen-hour journey. After looking for suitable terrain in Wales, Merrill landed the plane safely in a field at the village of Llwyncelyn.

"We picked a field where there were three cows," Richman told the *New York Times*. "Those cows sure were ladylike and knew good manners. They trotted off nicely as we came down and nobody minded."

A day later the *Lady Peace* arrived at Croydon Airport in London. The ping-pong balls drew the attention of British customs authorities, who debated whether to charge duty on the merchandise since the items were categorized as sporting goods. They finally agreed to accept the flyers' argument that the ping-pong balls were being carried as equipment and Richman was not compelled to pay import duties.

BELLANCA

A misunderstanding of local nomenclature complicated the rescue of Broadway singer Harry Richman's plane from a bog at Musgrave Harbour, Newfoundland. Left: A team, led by US First World War flying-ace, Eddie Rickenbacker (centre), mistakenly landed at Musgravetown, 125 kilometres away. Eventually, the *Lady Peace* – with Richman (right) and pilot Dick Merrill (left) – was extricated and flown to Harbour Grace (next page). Rickenbacker worked for Eastern Airlines, and flew one of the company's DC-2 aircraft, known collectively as The Great Silver Fleet. Above: Sheet music featuring Harry Richman.

Any notions the men had of refuelling the plane and immediately returning to New York had been dismissed by the time they reached London. The flyers were exhausted. Besides, Richman quickly became the toast of London high society, so the flamboyant showman decided to delay his return flight by a few days.

Inclement weather forced Richman and Merrill to extend their stay even longer. On September 14, after finally receiving reports of clear skies over the entire North Atlantic, the *Lady Peace* departed Southport, England, on a projected non-stop return trip to New York. The weather predictions were inaccurate. Stormy weather and dense fog drove the aircraft off course. With their fuel supply running low, Merrill was forced to make an emergency landing in Newfoundland. After fruitlessly searching for Harbour Grace airfield, he opted to land near Musgrave Harbour, a community on the northeast coast of Newfoundland.

The plane rolled along the grass before sinking into a peat bog. Although its nose tipped forward into the ground, the plane was largely undamaged. Fishermen mending their nets on shore hurried on foot over rough terrain to reach the aircraft. They raised the stricken machine from the marsh by sliding timbers under its wheels.

"That Newfoundland bog is sure soft stuff," Richman reportedly remarked as he drank tea before an open fire. "It sunk in like a feather bed, it seemed to me, and felt just about as good."

Upon learning of Richman's woes, Rickenbacker organized a rescue party. Two Eastern Air Lines planes left New York for Musgrave Harbour with mechanics and spare parts. But the Americans were unfamiliar with Newfoundland place names, and mistakenly landed at Musgravetown, about 120 kilometres south of Musgrave Harbour.

When Rickenbacker and his team finally arrived on the beach at Musgrave Harbour, they quickly repaired the *Lady Peace*. The machine was then dragged out of the mire onto a specially built wooden runway. Merrill and Richman flew to Harbour Grace, and then finished their journey to New York.

They arrived at Floyd Bennett Field on September 21. In spite of the extensive delays and numerous mishaps the flyers had achieved the first round-trip crossing of the Atlantic Ocean by plane.

The *Lady Peace* and Eddie Rickenbacker's DC-2 at Harbour Grace.

CHAPTER 8

Ditching at sea is an aviator's nightmare, and chances of survival are slim. Some, such as the crew of the *Esa* in 1931, were lucky to be found alive.

INCREDIBLE RESCUES

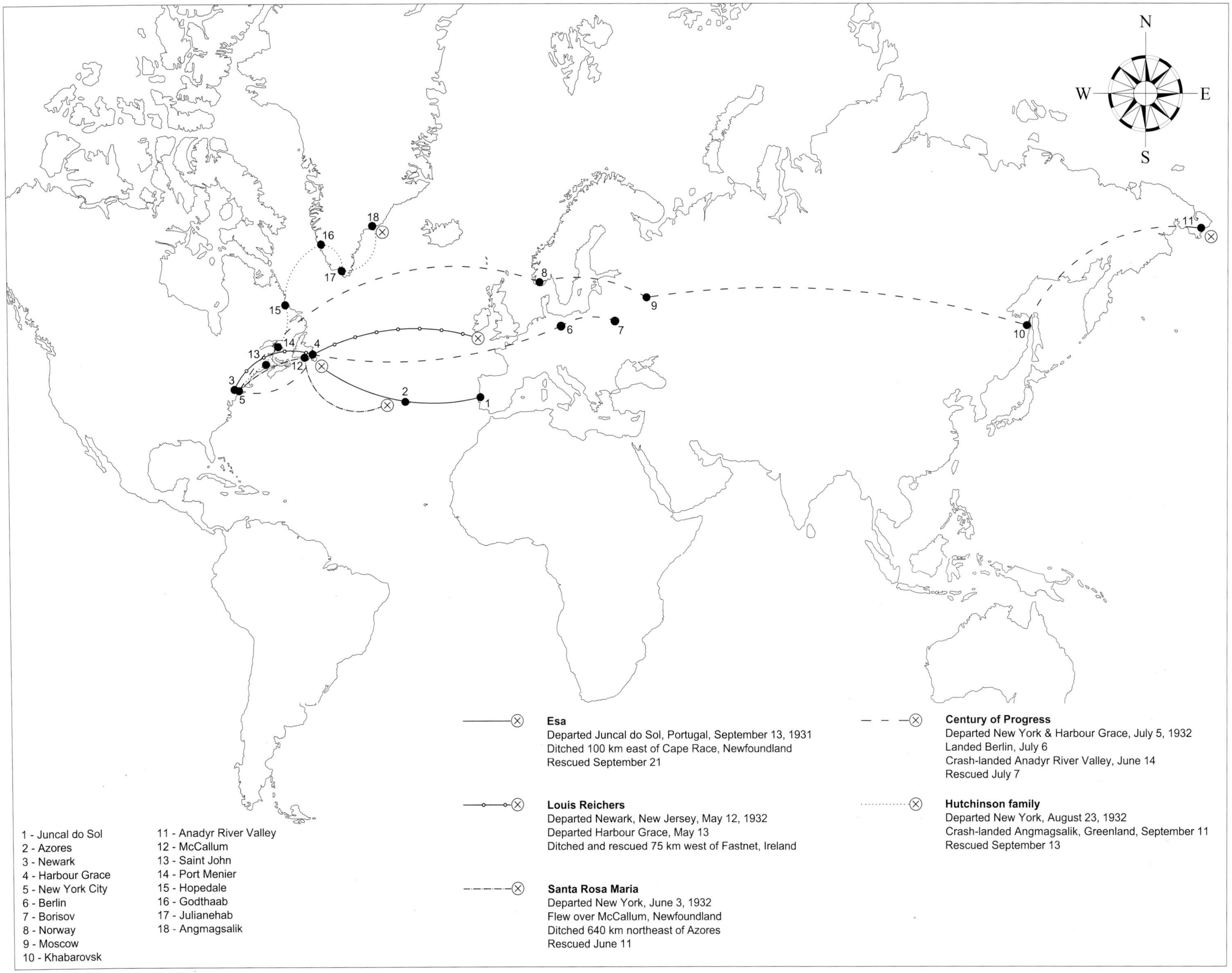

The history of aviation is marked by appalling disasters, but also by remarkable stories of courage and survival. Many unlucky or ill-prepared aviators perished during long-distance flights over the world's oceans, and only a small minority lived to retell their experiences after crashing or landing at sea.

Certainly, few people have survived an ordeal as lonely and painful as that endured by three men who took off on a transatlantic flight from Europe in 1931. A young German pilot, 21-year-old Wilhelm Rody, had been encouraged by the flight of the *Bremen* in 1928 to make his own westward crossing of the Atlantic. Rody planned to fly non-stop to New York, but also set an alternate route to Newfoundland if the weather deteriorated. After reaching New York, he intended to fly back to Europe.

He had used inheritance money to purchase a Junkers W-33 aircraft, and hired a former First World War flyer, Christian Johanssen, to be his co-pilot. A Portuguese passenger, Fernando da Costa Viega, accompanied them.

The plane was named *Esa*, after Rody's fiancée. Following a christening ceremony on September 13, 1931, at Juncal do Sol, Portugal, the *Esa* took off with Rody and his companions. Rody and Johanssen calculated that the *Esa* was carrying enough fuel to last 48 hours; the flyers also brought provisions: 12 bananas, 24 apples, sandwiches, and 3 loaves of bread. "Plenty," Rody said, prior to the flight, "to last us until we get there."

If the plane was forced to land on the ocean, Rody expected it could float for eight hours. Rather than carry a radio, the flyers opted to stow an additional drum of gasoline into the fuselage.

On the day it took off, the *Esa* was seen flying over the Azores and again on September 14 by a passenger liner, the *Pennland*, 130 kilometres east of Cape Race, Newfoundland.

This was the last sighting of the aircraft. Search planes were sent to scour the ocean, but their efforts were in vain. As the days passed, hope gave way to acceptance that the men were dead. So the world reacted with shock when, on September 21, all three were plucked alive from the plane's wreckage by the crew of a freighter. They were found 13 kilometres off the coast of Newfoundland.

Rody had vastly underestimated how long the aircraft could float. Rather than eight hours, they had been drifting helplessly in the ocean for more than six days on what remained of the plane's disintegrating fuselage. The men were in remarkably good physical condition, although Viega was suffering from exposure and a leg injury.

After spending six days marooned near the coast of Newfoundland with their disintegrating aircraft, a Soviet freighter rescued the three adventurers who flew the *Esa*.

In relating their ordeal, Rody said the flight had gone well until they passed the Azores. At that point strong headwinds started buffeting the plane, so he and Johanssen decided to fly to Newfoundland rather than continue on to New York.

But powerful winds continued to hamper the *Esa*. A cold front then moved in, and soon the pilots were battling a powerful storm. "Whirlwinds rocked us back and forth," Rody said in the *New York Times*. "They tossed us and battered at us so that we could not keep a straight course. Again and again we were hammered about in a circle."

Adding to their troubles, one of the six cylinders of the single-engine plane ceased working. The engine sputtered as the *Esa* struggled through the night. Worse yet, the engine would only function at full throttle, which meant that the plane was consuming fuel at an alarming rate.

In the morning the men were overjoyed to see the *Pennland*. Judging from the ship's course, Rody knew that the *Esa* was nearing land. The plane delighted passengers on board the *Pennland* by circling overhead. Despite the plane's mechanical problems, Rody said he and the other two men never considered ditching the *Esa* alongside the ship.

"We had had our chance of immediate safety, with that liner so close," Rody said. "But no [sic] one of us thought of our lives. We wanted to reach America in our plane and to land there."

So the *Esa* kept going – directly into the wind. Two hours later the fuel supply, which had lasted 36 rather than 48 hours, was gone. The engine coughed to a halt, and the *Esa* drifted towards the ocean waves.

"Our feeling was indescribable," Rody said. "Suspense tightened and twisted our nerves. We looked at each other. We looked at that ocean."

The plane dropped into the crest of a wave. Rody, Johanssen, and Viega opened the cabin door and scrambled onto the fuselage. They watched as the wings filled with water. Parts of the plane were ripped away by tumultuous ocean swells. The men lay on their stomachs, grasping desperately to the plane as the sea became rougher. Suddenly, a large wave demolished the cabin door, washing away all of their food and fresh water.

As night came, the men felt the cold eating away at them through their soaked clothing. They retreated to the cabin, only to find it half full of water. "Never shall we forget this night, or the nights that followed before our rescue," Rody said. "The first night we spent standing in rising water in the little cabin."

For the next six days, the men had only seven ounces of chocolate for food, and drank oily water from the engine's cooling system. As the days passed, the men grew weaker. Viega, whose leg had been injured as the plane was being tossed about in the air, became feverish. All of them found their skin peeling and their legs swelling from standing in water day after day. The plane was also disintegrating around them, and each day it sank a little lower beneath the water.

No ship passed them for over five days. Then, on September 20 they saw a freighter. "We waved our flag 'til all strength had left us," Rody recalled. "But that ship never saw us. It was unbelievable."

Despair set in. But the next day they spied another vessel. The men signalled again and watched apprehensively. It turned towards them; they'd been seen. "We were saved," Rody said.

The ship, a freighter carrying locomotives to the Soviet Union, transferred the airmen to a passenger liner heading to New York. Upon their arrival, Viega spent several days in hospital, where he made a complete recovery. Although Rody pledged to make another transatlantic attempt, there is no record of him doing so.

Unlike Rody and his companions, few of the aviators who failed in their attempts to cross the North Atlantic lived to recount their experiences. By 1932 twenty had died, with no bodies recovered and little wreckage found. Only the crew of the *Esa*, and Hawker and Mackenzie-Grieve in 1919, had survived a plunge into the icy waters between Newfoundland and Ireland – until 1932 when American air racer Louis Reichers set out from Harbour Grace.

Like many professional pilots of his era, Reichers financed his passion for flight by flying passenger planes and contracting his services to wealthy businessmen. One of his clients was Bernarr Macfadden, a controversial but successful health food and exercise guru. Macfadden, who had changed his name from Bernard MacFadden, also published sensational magazines, including *True Detective* and *True Story*.

Starting in 1930 Reichers began challenging speed and endurance records. After flying with Macfadden on a tour of Europe in 1931, Reichers was ready to make an attempt at breaking Lindbergh's record for flying from New York to Paris. Macfadden loaned Reichers one of his aircraft, a Lockheed Altair. Reichers changed its name from *Gold Eagle* to *Liberty* and had an image of the Statue of Liberty painted onto the plane's fuselage.

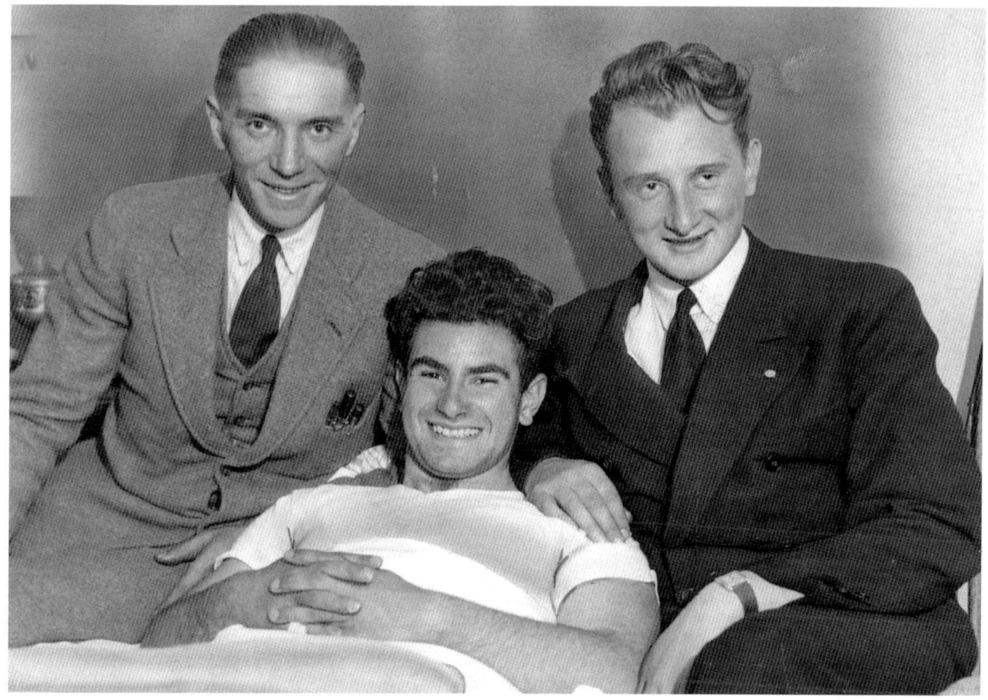

Despite their ordeal on the *Esa*, the crewmen were largely unharmed. From left: Christian Johanssen, Fernando da Costa Viega, and Wilhelm Rody.

Reichers left Newark, New Jersey, in the *Liberty* at 11:00 p.m. on May 12, and arrived in Harbour Grace in a record time of six hours and nineteen minutes. He refuelled the aircraft, repaired some minor damage to the plane's stabilizer, and, within a few hours, set out for Europe. Reichers planned to land at Baldonnel Airfield in Dublin before finishing his journey at Le Bourget Airport in Paris.

The first hour passed without incident, but, as he crossed the Grand Banks of Newfoundland, a thick layer of cloud enveloped the *Liberty*. After flying almost blind for more than ten hours, Reichers lost hope of reaching Ireland. As

After air racer Louis Reichers (above and next page) was forced to ditch his aircraft in the Atlantic Ocean, he was rescued by a passenger ship, the *President Roosevelt* (right). Three years earlier, in 1928, the liner had carried Amelia Earhart back to the US after she became the first woman to fly across the Atlantic.

night approached, and with little fuel remaining, he searched the ocean for the lights of a ship. He spotted a passenger liner, the *President Roosevelt*, and signalled the crew that he was about to ditch the plane and needed to be rescued. Reichers slowed the plane and landed with a jolt in the ocean.

"It was very dark and visibility so poor that I had to hang my head out to see when landing," Reichers wrote. "As I needed both hands to land, I could not protect my face, and the shock threw me forward, hitting the cowl; but outside of bending my nose out of shape – that required two stitches to straighten – there was no other damage."

Reichers had come tantalizingly close to success; he was rescued only 75 kilometres west of the Irish coast.

Reichers said Macfadden radioed him aboard the ship and reported that his sponsor was not concerned about losing the plane. After returning to the United States Reichers focused on being a test pilot, a risky occupation he had embarked upon two years earlier.

During the Second World War Reichers's experiences as a test pilot and long-distance flyer would prove valuable to the Allied forces. As a reserve pilot with the Army Air Corps, he was called up in 1941 to ferry bombers across the North Atlantic to England. The ferrying of aircraft by air from US factories to airfields in Great Britain had begun a year earlier. But in the fall of 1941 American commanders ordered a secret mission to determine the viability of establishing potential air routes across the Pacific Ocean.

On September 29 a B-24 Liberator bomber piloted by Reichers and Major Alva Harvey took off from Gander airport in Newfoundland with a crew consisting of personnel from the Army Air

Corps and Pan American Airways. After landing at Prestwick Airport in Scotland, the Liberator continued on to Moscow and from there to airfields along the Pacific rim.

The crew surveyed landing fields and assessed whether they would be capable of handling the large aircraft that would be needed in the event of a war with Japan. Reichers and Harvey returned to Washington DC on October 30, six weeks prior to the attack on Pearl Harbor. Reichers was then promoted to colonel and commanded the engineering section of Air Transport Command, which oversaw aircraft ferry operations for the US forces.

Brushes with death, such as those experienced by Reichers and Rody, were commonplace among pilots during the early decades of aviation. Most crashes were the result of poor pilot training, primitive navigational aids, and notoriously unreliable equipment.

Few pilots could count on the luck of James Mattern, an ambitious Texan who came close to losing his life in the Siberian Arctic. As a teenager, Mattern attended high school in Calgary, Alberta. Later, he spent several years in the US Army Air Corps before becoming an air racer and endurance flyer. In 1932 he prepared to better the round-the-world record of eight days set a year earlier by Wiley Post and Howard Gatty in the *Winnie Mae*. Another former Air Corps pilot, Bennett Griffin, agreed to be Mattern's co-pilot, and a pair of Oklahoma oilmen financed the flight.

Like Post and Amelia Earhart, Mattern and Griffin flew a single-engine Lockheed Vega. The plane, painted red, white, and blue, was named *Century of Progress*, in honour of the Chicago World's Fair which would take place in 1933. The fair, with its slogan A Century of Progress, celebrated the achievements of science and industry.

On July 5, 1932, Mattern and Griffin took off from Floyd Bennett Field, New York, on their world flight. Their first stop was Harbour Grace, but, as they flew over Newfoundland, the pilots became lost in a fog bank. When the fog cleared, they were flying over Cape Freels, a community on the coast of Bonavista Bay. Completely lost, the pilots wrote a note and dropped it to onlookers, asking for directions to Harbour Grace. Townspeople pointed them in the right direction, and four hours later they finally discovered the Harbour Grace airfield.

After a two-hour refuelling stop, they took off from Harbour Grace and flew directly to Berlin, rather than Ireland or England as other flyers had done. This put them three hours ahead of the record time set by the *Winnie Mae*. Mattern and Griffin left for Moscow the next morning – they were confident that the *Century of Progress* would smash the world flight record.

At this point their luck ran out. The plane's engine failed over the Ukraine, forcing the pilots to land. But what they presumed was solid turf turned out to be a peat bog. The landing gear sank, pitching the nose of the plane forward and into the ground. Mattern and Griffin were only slightly injured, but the plane had been badly damaged. The men were listed as missing for two days, until Soviet officials disclosed that they had crashed at Borisov, a town near Minsk.

Despite this setback, Mattern refused to give up. On June 3, 1933, he again took off from Floyd Bennett Field – solo this time – flying the *Century of Progress* on another round-the-world flight. He planned to stop at Chicago towards the end of the journey to celebrate the world's fair. Rather than refuelling in Newfoundland, he continued non-stop to Europe. Nothing was heard from the *Century of Progress* until word came the next day that Mattern had become lost and had landed on a rocky beach in Norway. Despite losing his way, he still managed to reach Moscow several hours ahead of the time set

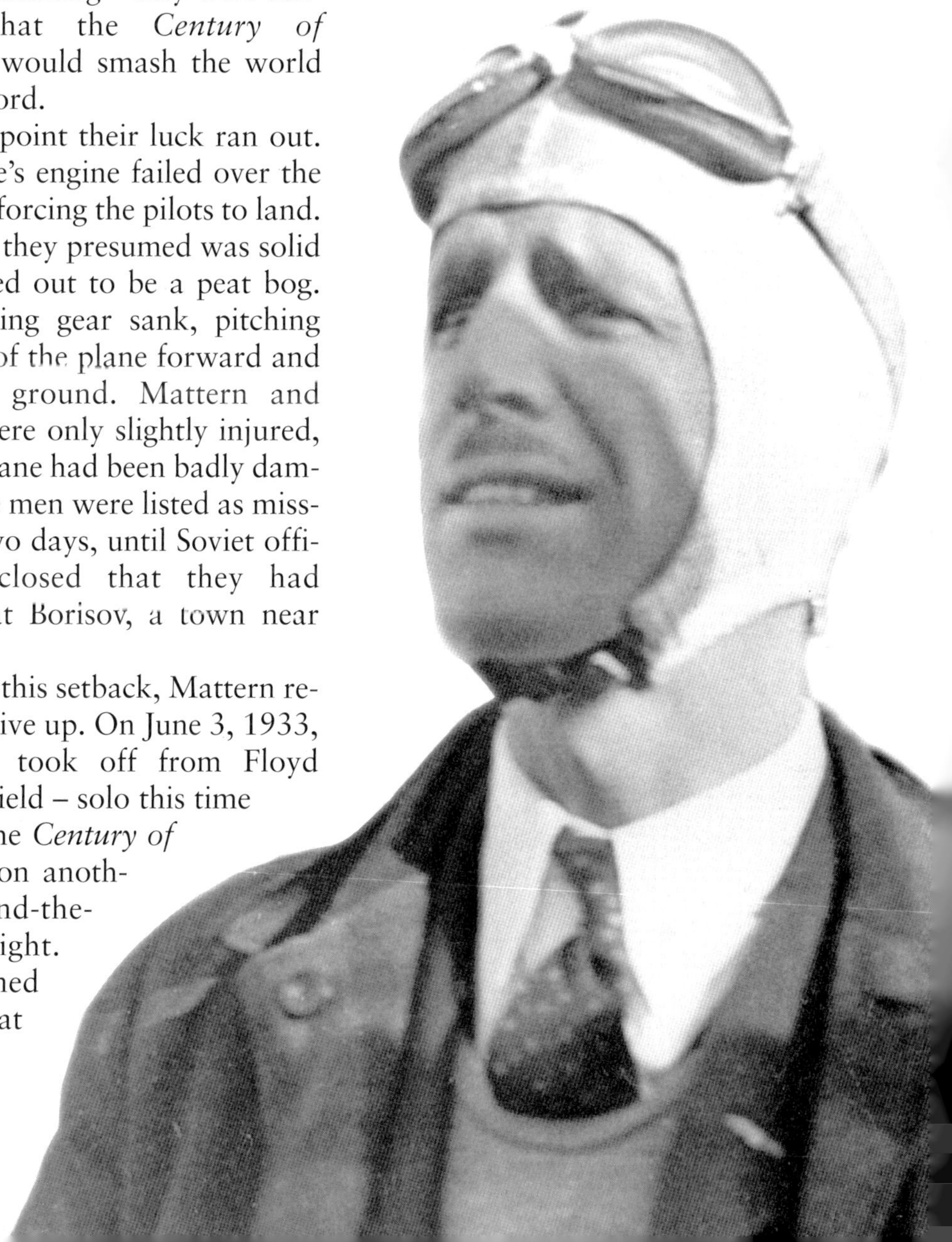

by the *Winnie Mae* in its 1931 flight.

But Mattern's troubles were only beginning; mechanical problems continued to plague the *Century of Progress* as it flew across the Soviet Union. On June 14 Mattern took off from Siberia for the last time and headed for Nome, Alaska. He never arrived. As the days – and then weeks – passed, hope faded. He was presumed dead until news broke, on July 7, that he was alive, after being missing for 23 days.

His story of those three weeks was one of pain, loneliness, and determination. It was also a tale of generosity and caring from those who saved his life.

The *Century of Progress* had flown without trouble for several hours after taking off from the city of Khabarovsk and flying along the Sea of Okhostk. But the weather deteriorated, so Mattern flew inland along the Anadyr River Valley. He decided to land at the town of Anadyr, before heading over the Bering Sea to Alaska.

Suddenly the oil level dropped, the engine began overheating, and Mattern was forced to land on the rocky tundra. As its landing gear struck the ground, the plane catapulted forward, smashing a wing and crushing the fuselage.

Mattern lay trapped in the cockpit for 24 hours, nursing a badly twisted ankle and bruises. Finally, he summoned the strength to smash a hole through the cabin with an axe. He lived in the rear of the plane for another eight days, with only eight chocolate bars to sustain him.

Once his rations had been eaten Mattern decided to walk to the Anadyr River, where he hoped a boat might pass by and rescue him. Carrying a rifle and fishing tackle, he hobbled to the river – his ankle throbbing in agony. On his second day at the river Mattern saw two boats. He fired several shots, but nobody noticed.

The cold was excruciating, so he attempted to build a shelter. Using spruce boughs, he managed to build a small makeshift tent, but it failed to keep the wind out. So Mattern abandoned it and, with great effort, chipped a hole in a nearby glacier, crawled into it and slept.

He then decided that the best place to be noticed was on an island in the middle of the river. He built a raft, but it flipped over – tossing him into the water. After struggling back to the riverbank, he started a large fire. "When I took off my flying suit, my underclothing, which had been soaked with gasoline from the crash, caught fire, and I had to jump back into the icy river," Mattern recalled.

The *Century of Progress*, named in honour of the 1933 Chicago World's Fair, flying over New York City (right). A crowd gathers in Harbour Grace (above) as James Mattern and Bennett Griffin prepare to cross the Atlantic. Below left: The World's Fair was celebrated by the US Postal Service, which produced specially stamped envelopes for the event.

A second attempt at building a raft proved successful. He got to the island and waited for help – his strength sapped. It was then that a group of Inuit men found Mattern; they took him to their camp, where he was nursed back to health.

For the next week Mattern observed the lives of the nomadic families who tended to him. He observed their work, entertainment, and rituals. "A caribou-killing ceremony and feast were held," Mattern wrote in the *New York Times*. "After the feast the tent was darkened. It was already smoke-filled. A drum was produced, and we sat there for five hours, listening to the same monotonous beat and to the cries of the natives until they had worked themselves almost into hysterics."

A few men agreed to take him by boat to the town of Anadyr. "For the next two days every stroke of the oars sang out the grand old word – 'America'," he wrote.

Mattern remained in Anadyr for more than a week, until the Soviet government could find a plane to fly him to Nome, Alaska. However, bad luck would continue to dog Mattern during his return trip to the US. First, the Soviet Sikorsky flying boat ran out of fuel before reaching land,

six kilometres from Alaska – Mattern and some of the crew were compelled to row ashore in a rubber raft. He was then flown to Fairbanks, and to several small towns in British Columbia.

Another pilot flew him east, with Mattern planning to fly solo to New York from Toronto. But the plane was forced to make a crash landing in northern Ontario, so he continued by car to Toronto. He finally arrived at the Floyd Bennett Field in a borrowed plane on July 30.

By then, Mattern had been largely forgotten, as Wiley Post had stolen the headlines once again. During Mattern's struggle for survival in Siberia, Post had thrilled the American public by breaking his own record for flying around the world.

In 1932, the year Mattern made his first attempt at circumnavigating the globe, another would-be hero was planning his own transatlantic heroics. When Stanislaus Hausner embarked up-

Martha Hausner bids farewell to her husband, Stanislaus, as he departed New York on a solo attempt at reaching Warsaw, Poland. Hausner was forced to ditch his plane in the ocean, and waited eight days to be rescued.

on his solo flight across the Atlantic in 1932, he expected to become the first person to fly non-stop to Warsaw, Poland, from New York. Instead, Hausner would set a different kind of record – for time spent marooned at sea in a downed aircraft.

Hausner was a native of Poland who, as a child, moved to New Jersey with his parents. Being deeply religious, he joined a Catholic monastic order. He resigned the order in 1920 and worked for several years as a motion picture film operator. He also fell in love with aviation, and in 1929 he became a pilot.

In 1932, despite having little experience as an aviator, Hausner decided to honour the nation of his birth by making a non-stop flight to Warsaw. He acquired a Bellanca monoplane and named it *Santa Rosa Maria*.

Despite the mounting death toll among aviators who had attempted to fly across the Atlantic, he did not seem to fully appreciate the dangers. In late May 1932 he took off from Floyd Bennett Field, New York, and hoped to reach Europe within 24 hours. But Hausner did not heed the forecasts of poor weather over the Atlantic. After battling fog and cloudy conditions for six hours, he turned around and landed safely in New York.

On June 3, he tried again, even though weather forecasts showed dense fog and low temperatures east of Newfoundland. At the airport that morning, his wife, Martha, helped him stow provisions – ham and chicken sandwiches, oranges, apples, coffee, and water.

"Well, I guess it's time to go," he said, before kissing Martha and taking off. She and Father Paul Knappek of St. Casimir Roman Catholic Church in Newark, New Jersey, followed the *Santa Rosa Maria* in another aircraft. After accompanying Hausner for an hour, Martha and Father Knappek returned to Newark; they went immediately to church and prayed.

Hausner would soon need all the help – divine or otherwise – he could muster, although the first leg of the journey passed with relative ease. After twelve hours of flight, the *Santa Rosa Maria* was seen over McCallum, a community on Newfoundland's south coast. But as he continued flying east the aviator ran into an intense low-pressure weather system. Strong winds slowed the plane and dense fog impaired navigation.

Airports in Ireland, England, and Poland were put on alert for the plane. But nothing was seen of it. As the days passed, practically all hope was lost for Hausner's survival. Only his immediate family and Father Knappek had faith that he remained alive.

Their confidence was rewarded on June 11, eight days after Hausner had departed on his transatlantic journey. A British oil tanker, the *Circe Shell*, had plucked the aviator from his partially submerged aircraft. It was floating, tail up, 640 kilometres northeast of the Azores. "There, I told you he was too good a pilot to have been killed," Martha Hausner told the newspapers.

Upon being rescued, Hausner was suffering from exhaustion, hunger, and dehydration, but he had fully recovered by the time the *Circe Shell* reached Miami, Florida, on June 22. The plane was salvaged by another ship a month later.

Even though he had not come close to achieving his goal, Hausner became a hero to many Polish Americans. A parade was held for him in Newark, and large crowds greeted him at several official receptions.

Despite intense public interest in his tale of survival, he refused to discuss his ordeal with the newspapers. He simply reported that fifteen ships had passed by before he was finally rescued. Hausner intended to write a book based upon a diary he had kept while clinging to his stricken aircraft.

But fate would intervene before Hausner's book was completed. In 1935 he was ready to make another attempt at flying to Warsaw, this time starting from Detroit, Michigan. A group of Polish Americans from Detroit had raised money to purchase an aircraft; it was named *Marshall Pilsudski* in honour of a visit to the US by Polish President Jozef Pilsudski.

On May 18, 1935, one month before Hausner expected to fly across the Atlantic, he took off from Detroit's Wayne County Airport. He flew over a mass being held for Pilsudski outside Sweetest Heart of Mary Roman Catholic Church in Detroit.

As Hausner circled over the congregation, a wing suddenly sheared away from the aircraft. More than 6,000 people watched in horror as the plane plunged earthward. It exploded into a fireball upon plowing into the backyard of a home.

"Hausner was circling high overhead, with the belfry of the church as the centre of the spirals," recalled a priest who witnessed the event. "He apparently made a power dive to gain impetus, and as he came out of it, I saw the right wing crumple off against the sky. The plane dropped like a plummet. There was a muffled explosion and a bright flash of flame. By the time I had run to the scene the plane was almost completely consumed by fire." A bugler, who had viewed the accident from the church steps, played taps as smoke billowed from the wreckage.

Such stories of triumph and failure among the aviation fraternity filled the newspapers and newsreels of the post-First World War years, but few used the media to promote their activities as successfully as "The Flying Family," whose adventures captivated radio listeners in the early 1930s.

"The Flying Family" had become highly popular by the time they embarked upon a transatlantic attempt in 1932. The crew and passengers included (seated from left) film director Norman Alley, and the Hutchinsons: George, Janet Lee, Blanche, and Kathryn. Standing (from left): navigator Peter Redpath, engineer Joseph Ruff, and radio operator Jerry Altfilisch.

The escapades of George and Blanche Hutchinson and their two young daughters, Janet Lee and Kathryn, provided them with a steady income for several years during the Great Depression. However, one of the Hutchinson family's adventures would end with a desperate struggle for survival on the desolate, rugged coast of Greenland.

The family was led by George, who harboured two passions: gambling and aircraft. In the mid-1920s, after winning US $8,000 from betting on a horse race, he purchased his first plane, and then learned how to fly. He managed – and subsequently purchased – an airport, and earned money by flying people on short excursions.

For Hutchinson, aviation became an obsession. "[Flying] became a disease; it's all he could think about," Janet Lee Hutchinson recalled.

The Hutchinsons first attracted public attention in 1929 when George and his brother Leonard flew with Blanche and the girls in an Army biplane. This stunt marked the beginning of a lucrative flying career for the family, which would end with the outbreak of war in 1939.

In 1931 "The Flying Family" embarked on an air tour of all 48 states in the US mainland, making regular radio broadcasts throughout their journey. The purpose of this, and their transatlantic attempt a year later, the daughters said, was to prove that air travel was not the preserve of the brave or the rich.

"Lindbergh was pushing flying the mail, and he thought that was the greatest asset of aviation," Kathryn Hutchinson said. "Daddy saw this as being secondary to getting people – families – to accepting aviation as a means of transportation. That's why he took the family with him."

With the backing of US President Herbert Hoover, the Hutchinsons succeeded in visiting each of the 48 states. At each stop they delighted spectators with their mascot, a lion cub named Governor, in honour of New York Governor Franklin D. Roosevelt, who joined Hoover in supporting their efforts. The family's progress was closely followed by hundreds of thousands of Americans, many of whom listened to their radio broadcasts and telephoned to receive a commemorative jigsaw puzzle offered by the Hutchinsons. "We were on the radio – people called in and got pieces of [a special] jigsaw puzzle," said Kathryn, who was seven years old at the time. "Over a million people phoned in."

Throughout the journey, George and Blanche kept the girls busy by looking out for landmarks. "You flew down low and read the name of the town on the railroad station or the water tower – there were no accurate maps available," Kathryn said. "Daddy would say, 'This is what you're going to look

Kathryn (left) and Janet Lee Hutchinson (right) received much of their schooling in the skies over North America. They, together with their mother, Blanche (centre) and father George, comprised "The Flying Family."

for'. He'd tell us the letters and we'd go down lower and find out where we were."

Their only mishap occurred when George opted to cross into British Columbia from Oregon to avoid bad weather. But instead of missing the storm, they flew straight into it while crossing over the Rocky Mountains. The aircraft crash-landed, forcing the family to delay completion of the tour until repairs had been made.

Despite this accident, the family's cross-country tour had been a success. Emboldened by the accolades, George and Blanche decided to attempt an even more ambitious project: crossing the Atlantic Ocean.

Using a newly purchased twin-engine Sikorsky aircraft, the family took off from Floyd Bennett Field, New York, on August 23, 1932 – Kathryn and Janet Lee were eight and six years old at the time. The plane was named *City of Richmond*, after Richmond, Virginia, where George had spent much of his childhood.

Joining the Hutchinsons were three crewmen and a passenger: Canadian navigator Peter Redpath, mechanic Joseph Ruff, radio operator Jerry Altfilisch, and film director Norman Alley.

The eight adventurers landed first in Saint John, New Brunswick, where the aircraft's wheels were replaced with pontoons. The following day they flew to Port Menier, a resort community on Anticosti Island in the Gulf of St. Lawrence. Here, they stayed at a luxurious hunting lodge overlooking the St. Lawrence River.

On August 30, after waiting several days for a favourable weather forecast, they left Anticosti Island. The aircraft cruised along the coast of Quebec and Labrador, before reaching the community of Hopedale. In his book, *The Flying Family in Greenland*, George Hutchinson describes their arrival:

> From the air we could make out the red-and-white building belonging to the Hudson's Bay Company, and used as a general store. Close by we saw the yellow [Moravian] mission house, the church, and the radio station with its two high towers. Looking more closely as we swooped down, we saw many husky dogs lying outside on the ground, among the dilapidated, unpainted wooden huts.
>
> As we circled the village it suddenly came to life. Out of doors popped Eskimos of all ages, and came running down to the water's edge, with the dogs barking and running in all directions. The Eskimos grabbed our hands and shook them in a gesture of welcome, while wide smiles spread across their faces.

The visitors stayed at the Moravian missionary residence, home to Reverend Walter Perret and his family. Hutchinson described the poverty of the Inuit in the community, and criticized the Moravians for encouraging the Inuit to rely upon the missionaries for food, clothing, and medical care.

> Perritt [sic] told us of the work of the Moravian mission, in providing for and educating the Eskimos. We all listened attentively to him of how poor [the Inuit] were in Hopedale. But all this has come about since the white man has attempted to civilize the native. He has made the Eskimo dependent.

On September 2 the visitors took off for Greenland, arriving in Godthaab on the west coast of the Danish territory. When the *City of Richmond* arrived, the Hutchinsons did not receive the warm welcome they had come to expect in their travels. Denmark had refused to grant them landing rights in Greenland, even though permission was given for two other Danish territories, Iceland and the Faroe Islands.

The Hutchinsons were fined the equivalent of US $180. More importantly, they were also compelled to change course; rather than fly to the east coast town of Angmagsalik by crossing the Greenland ice cap – a journey of 740 kilometres – they were ordered to follow the coastline, a distance of almost 2,000 kilometres. The authorities reasoned that the coastal route was safer than flying overland, an argument George Hutchinson dismissed. The order would have disastrous consequences for the Hutchinsons and their crew.

After a delay of several days, the *City of Richmond* left Godthaab. The plane reached the coastal town of Julianehab, near the southernmost point in Greenland. Here, they waited for favourable weather before leaving for Angmagsalik.

They departed on September 11, but the *City of Richmond* never arrived in Angmagsalik. As the plane neared the community, ice built up along its wings as it flew through dense cloud cover. Thickening ice forced the plane into a slow but inexorable descent, and the crew had no choice but to land in the ocean. Altfilisch, the radio operator, managed to send out an SOS before water swamped the batteries and cut electricity.

The plane taxied to a small island, just 55 kilometres from Angmagsalik. Passengers and crew hastily unloaded food, clothing, guns, cushions, tools, and the radio. They also tore a large section of canvas from the wings to build a shelter. As they worked, waves pounded relentlessly against the aircraft. In his book, Hutchinson described the final moments of the *City of Richmond*:

> Two 'bergs were gradually drifting closer and closer to the plane as we worked. I don't know exactly how long we worked, but I was the last on board, when I felt the vessel shudder and begin to crumble. I jumped ashore, and we all stood there helplessly, watching the two remorseless icebergs crush the *City of Richmond* between them, and pull this fine, sturdy 'plane under the waters of the cold Atlantic.

The castaways built a makeshift shelter by pulling the canvas between several large rocks. Inside, they laid out rugs and cushions, and in a corner built a rock fireplace. With no trees on the island, they lighted cans of grease to heat themselves and the food supplies taken from the plane.

They cooked a 17-kilogram lamb that the Hutchinsons had intended to deliver to the governor of Angmagsalik as a gift from the governor of Julianehab. "We used these big cans of lubricant and had greased smoked lamb," Kathryn Hutchinson recalled.

In his account, George described the ordeal as being little more than an adventure. Reality, however, was different. Altfilisch managed to get the radio working to receive messages, but he was unable to broadcast. The group huddled inside the makeshift shelter, listening helplessly to radio messages from search vessels.

"I wasn't sad or nervous until I saw Daddy crying," Kathryn said. "We could hear on the radio that they were looking for us, but they weren't really headed in the right direction."

Two days after being marooned, and with the fuel supply running out, the crew of a British fishing trawler, the *Lord Talbott*, spotted their fire. Using the vessel's lights, a Morse code message was sent to the island, asking them to build a second fire if they were indeed the Hutchinson family.

"By then we had used all of the grease except for this last can that was going," Kathryn said. "We had to light something. One of the men [Norman Alley] was a photographer who was taking movies. [The film] was all we had left to burn. That was really too bad."

As the precious visual account of their journey flared into the night sky, the trawler sent a smaller boat to rescue the Hutchinsons and their crew. Despite the ordeal, the Hutchinsons continued their journey, travelling with the *Lord Talbott* to Iceland and Scotland. A parade was held in their honour in Edinbugh, and they were fêted in London.

Joseph Kennedy, US ambassador to Great Britain, invited them

Kayakers get a close look at the *City of Richmond*, upon its arrival in Hopedale, Labrador.

to a reception at his residence – an honour George Hutchinson rebuffed. "He wouldn't go because he didn't approve of Joe Kennedy," Kathryn said. "He started his fortune with bootlegging. Daddy was dead set against alcohol."

"The Flying Family" returned home by ship. Seven years later, in 1939, they set off on another ambitious flight – to visit every nation in the world. This was a goodwill mission, designed to promote peace. They brought with them a scroll, which was to be signed by leaders from all countries.

They reached every nation in North America, South America, and the West Indies. By then, however, war had broken out in Europe, and the Hutchinsons were forced to halt their tour.

CHAPTER 9

Transatlantic passenger flights were close to becoming reality in the late 1930s, following trials by Imperial Airways and Pan American World Airways. One of Imperial's test planes, the *Caledonia*, is seen above New York City in 1937.

A SHRINKING PLANET

1 - New York City
2 - Paris
3 - Bangor, Maine
4 - Gander
5 - Old Orchard Beach
6 - Burgeo
7 - Foynes
8 - Botwood
9 - Montreal
10 - Shediac

Caledonia
Departed Foynes, Ireland, July 5, 1937
Landed Botwood, July 6
Arrived Montreal, July 8 and New York, July 9

Clipper III
Departed Botwood, July 5
Landed Foynes, July 6

Howard Hughes
Departed New York & landed Paris, July 10, 1938
Circumnavigated the world; arrived New York, July 14

"Piggy Back"
Departed Foynes, July 20, 1938
Landed Montreal, July 21

Carl Backman
Departed Bangor, Maine, & Gander, Newfoundland, May 15, 1939
Disappeared (1 of 1 dead)

Thomas Smith
Departed Old Orchard Beach, Maine, May 28, 1939
Crashed near Burgeo, Newfoundland (1 of 1 dead)

By 1938, the wild, youthful years of aviation had almost come to a close. Two events – a new round-the-world speed record, and stringent government regulation of aircraft and pilots – marked the beginning of a new era, one where members of the general public could buy a ticket and fly almost anywhere in the world.

It was perhaps fitting that Howard Hughes would end the pioneer phase of flight. Although he had inherited a drilling tool company, Hughes demonstrated little interest in the firm; instead the eccentric millionaire sustained a lifelong obsession with flight.

Hughes used the profits earned by his businesses to feed his love for aviation, as well as his other passions, principally motion pictures and beautiful film starlets. These three interests sometimes converged; as a movie producer he released *Hell's Angels*, a 1930 film portraying aerial battles of the First World War. The production, which starred Jean Harlow, became the first multi-million-dollar movie, costing Hughes US $3.8 million. Hughes, who also directed the film when the original director resigned, hired former First World War pilots to fly stunt planes, three of whom died while flying battle scenes.

Hughes was also a pilot, and he revelled in setting speed records. Although he was a wealthy man who could easily hire pilots for dangerous flights, Hughes insisted on personally test-flying new aircraft – a dangerous occupation at a time when aeronautical engineering was in its infancy. Hughes was badly injured on one occasion when the prototype of a plane, the XF-11, which was designed and built by his company, malfunctioned during a test flight and crashed.

Like other aviators, most notably Charles Lindbergh, Hughes also played a central role in designing aircraft. Despite having no formal education as an engineer, Hughes improvised in order to improve the speed and efficiency of his planes. In 1935 he set a US coast-to-coast speed record in a new aircraft, the H-1. This engineering prototype would influence the design of Second World War fighter planes, with its drooping ailerons, split flaps, hydraulic landing gear, fire-suppression system, and movable pilot's seat.

In 1938, he decided to smash the record for flying around the world set five years earlier by Wiley Post. Hughes reputedly spent approximately US $300,000 financing the venture, which included buying a specially built Lockheed twin-engine aircraft. He named the plane *New York World's Fair 1939* in order to promote ticket sales for the event.

But there was a second reason for the urgency of this flight. Despite Hughes's bravado, time was rapidly running out for the free-spirited men and women who rode their machines into the skies. In February 1938, legislation had

Millionaire and legendary eccentric, Howard Hughes, was obsessed with flight. In 1938, he smashed the round-the-world record set by Wiley Post.

been introduced to the US House of Representatives that would establish an independent agency with the authority to regulate air traffic. Among its powers, the new body would certify pilots, aircraft, and airfields. It could also ban so-called stunt flights.

By June the new regulatory agency, the Civil Aeronautics Authority, had been approved by Congress and was signed into law by the president. But it would not actually take effect until early August.

The Civil Aeronautics Authority would soon have the ability to disallow Hughes's world flight, so he dedicated himself to completing the journey before its regulations came into effect. On July 10 the *New York World's Fair 1939* took off from Floyd Bennett Field in New York. Besides Hughes, the aircraft held a crew of four men. One of his co-pilots was Harry Connor, who had flown across the Atlantic Ocean with Canadian Erroll Boyd in 1930. Other crewmen included co-pilot and navigator Thomas Thurlow, engineer Edward Lund, and radio technician Richard Stoddart.

Poor weather over Newfoundland waters slowed their progress across the Atlantic, and Hughes worried that the plane might run out of fuel before reaching Europe.

The crew conserved fuel by reducing speed. Despite slowing down, the aircraft reached Paris in 16 hours, 35 minutes – half the time Lindbergh took in 1927.

The record-setting pace continued unabated as Hughes flew across Europe, the Soviet Union, and then back to North America. When Hughes returned to Floyd Bennett Field on July 14, his time of three days and nineteen hours had halved Post's record.

A crowd of 20,000 people mobbed the often-reclusive millionaire as he stepped from the plane. The next day an estimated 1.5 million people lined the streets of New York in a tickertape parade held to honour his achievement. When the festivities ended, sanitation workers had to clean 1,800 tonnes of trash from city sidewalks.

Perhaps wisely, in light of the new aviation laws coming into effect, Hughes refused to take full credit for the achievement. Instead, he characterized the flight as a work of engineering prowess, rather than a successful conclusion to a risky adventure. This assessment was supported by the newly appointed members of the Civil Aeronautics Authority, who congratulated Hughes for his "most effective demonstration possible of the contribution that scientific preparation and organization can make to safety and speed in the air."

One of Hughes's most important contributions to aviation would be the creation of a passenger airline, Trans World Airlines, which began competing for US and transatlantic passenger traffic starting in the 1940s.

Hughes's legacy would continue well beyond his death in 1978. Hughes Aircraft Company, which had been established in 1935, became a world leader in high technology systems for scientific, military, and global applications. Raytheon Company acquired the firm in 1998.

The Lockheed Super Electra piloted by Howard Hughes above Manhattan on a test run prior to his round-the-world flight in 1938.

Hughes's round-the-world flight in 1938 marked the end of the pioneering era of flight. The aviation industry was maturing, with airline companies establishing passenger and airmail routes across North America and within Europe. In addition, large flying boats were beginning to prove their reliability over the world's oceans.

People were becoming more interested in buying airline tickets and experiencing the convenience of flying than in following the aerial feats of intrepid pilots. But this did not stop two more adventurers from trying their luck in crossing the Atlantic prior to the advent of the Second World War. Their experiences demonstrated that ocean flights were still dangerous for aviators and could claim the lives of the foolhardy and the unlucky.

The newly built Gander airport received its first aircraft from outside Newfoundland in May 1939 when Carl Backman arrived from Bangor, Maine. Backman, a pilot of Swedish heritage, was planning to fly non-stop to Stockholm.

To circumvent US authorities, Backman found a loophole in the regulations that effectively banned transatlantic flights by small aircraft. He claimed to have sold his plane to a Swedish buyer and alleged that he was merely delivering the craft to Sweden.

The 90-horsepower, single-engine monoplane carried no radio, and Backman brought only sandwiches and a thermos of tea with him as he took off on May 15 for Europe. He was never seen again.

On May 28, another lone flyer, Thomas Smith, took to the air from Old Orchard Beach, Maine, in a small plane, an Aeronca 65-C, which he named *Baby Clipper*. Fully loaded, the aircraft weighed just 860 kilograms.

Smith, who said he wanted to demonstrate the practicality of using light planes for long-distance flights, expected to reach England within two days. The *New York Times* reported that Smith had spent just $30 on fuel for the entire 4,800-kilometre journey.

Like Backman, Smith did not bring a radio for the flight, although he carried a first-aid kit, a revolver, and mosquito netting in case the plane was forced down over land. After taking off, Smith disappeared. His fate remained unknown until 1941, when a Canadian Air Force patrol aircraft discovered the *Baby Clipper* near Burgeo on Newfoundland's southwest coast. Smith's body was never found, but inside the cockpit was a note written by him, which indicated that he had survived the crash and attempted to walk to safety:

> Have some food and emergency supplies. I'm walking South, then will walk west if I hit ocean. North-

A Pan American World Airways flying boat, the *Clipper III*, at Botwood (left) in 1937, proving that transatlantic passenger flights were close to reality. It is being serviced (above).

Northwest. That is down the mountain. If I can't find a house will try to come back to ship. Weather sleeting – Thermometer dropping – Am afraid to stay in ship for fear of freezing while asleep.

In a rare partnership between rival airlines, Imperial Airways and Pan American conducted joint test flights of their flying boats on the treacherous North Atlantic route. Imperial had to strip the *Caledonia* (above, in Botwood) of all non-essential equipment in order to fly non-stop across the ocean in 1937.

While the short-lived age of air racing was coming to a close, a new era of commercial aviation was already maturing. Airline companies were now looking to bridge the world's oceans, this time using planes rather than airships such as the *Hindenburg*.

On July 5, 1937, a group of spectators watched in awe as a huge, four-engine Pan American World Airways flying boat skimmed along the waters of the Bay of Exploits and lumbered into the sky.

The Sikorsky S-42B, named *Clipper III*, was leaving Botwood and heading towards the sea base town of Foynes in Ireland. Unlike previous transatlantic attempts, this flight was not being conducted by a group of thrill-seekers and there would be no cash reward waiting for Captain Harold Gray and his three crewmen.

Instead, this was a rare collaboration between two rival airlines, Pan American and Britain's Imperial Airways. They were hoping to establish commercial transatlantic air services linking Britain, Ireland, Newfoundland, Canada, and the United States.

When the *Clipper III* disappeared over the horizon in central Newfoundland, the dignitaries visiting Botwood for the event did not return to their homes. Instead, they waited in anxious anticipation for the second partner in this unique transatlantic experiment.

They would not be disappointed. The next morning, July 6, another aircraft appeared on the horizon. With its engines roaring, an Imperial Airways flying boat, a Short Empire Class craft named *Caledonia*, landed in the Bay of Exploits and cruised towards land. It had taken the aircraft sixteen hours to fly to Botwood from Ireland.

Waiting on shore to greet Captain Arthur Wilcockson and his three crewmen was Newfoundland's governor Humphrey Walwyn who had travelled from St. John's to watch the two aircraft achieve this milestone in aviation history. By the time the *Caledonia* had reached its destination, the *Clipper III* had already arrived safely at Foynes.

The *Caledonia* flew to Montreal on July 8 and completed its journey by arriving in New York the next day. It made a return flight to Ireland a week later. The *Clipper III*, meanwhile, flew from Foynes to Southampton, England, before returning to New York, via Botwood and Shediac, New Brunswick. Imperial Airways and Pan American aircraft passed through Botwood and Shediac more than a dozen times during this first year of transatlantic flights.

The timing of these trial flights was no accident. Just one year earlier, in 1936, the *Hindenburg* had exploded while landing in New Jersey, a catastrophe that put an end to passenger airships, which at that time offered the only transatlantic air service. Despite the disaster, there was a growing public demand for a fast, convenient form of air travel, and the void left by the German airships provided an opportunity for Pan American and Imperial Airways.

These two companies were in an ideal position to take charge. Several years earlier, in 1933, they had begun planning a transatlantic flying boat service. This collaboration between the airlines also involved the governments of the United States, Canada, Newfoundland, and Britain. Botwood was selected as an operational base following an extensive survey in 1933 by Charles Lindbergh and Anne Morrow Lindbergh, who were hired by Pan American to survey possible landing sites in North America and Europe.

Lindbergh had long believed that land planes would eventually supplant flying boats as the most effective method for crossing the Atlantic, principally because sea bases needed sheltered harbours,

A refuelling barge and flying boat station were built at Botwood by Pan American and Imperial Airways. Similar facilities had also been installed in England, Ireland, New Brunswick, and Quebec.

and they were often closed by poor weather. Once paved runways became commonplace, Lindbergh expected commercial air carriers to switch to land planes for ocean crossings. But during the 1930s the world's major airlines focused on flying boats, primarily due to the scarcity of paved airfields. Seaplane terminals were also significantly cheaper and faster to build than airports.

Botwood was a logical place to situate a seaplane base, because the community is located at the end of a long, narrow fjord. The waters are generally calm, although sea ice during the winter and spring limited the flying season to only five months of the year.

Botwood was already well known to aviators. For several years, beginning in 1920, pilot Sidney Cotton used Botwood as a base for his planes. In addition to delivering mail to isolated communities, Cotton had been hired to spot seal herds on pack ice during the annual Newfoundland seal hunt. The Australian-born airman left Botwood following a contractual dispute with the Newfoundland government in the mid-1920s.

After winning exclusive rights to use Botwood for transatlantic flights in 1933, Imperial Airways signed an agreement with Pan American that allowed its US counterpart to share the future seaplane base.

Although flying boats were new to Newfoundland, by 1937 they had become a fairly common form of transportation in many parts of the world. The aircraft were romantically dubbed "clippers of the skies," akin to the swift, sleek clipper sailing vessels of the nineteenth century.

The transatlantic trial flights of 1937 were designed to prove the viability of flying seaplanes across the ocean on a commercial basis. To the public, the experiment appeared to be an overwhelming success. But the airlines knew otherwise; in fact the flights proved that improvements in engine power and fuel efficiency were required. Existing aircraft were not yet powerful enough to carry heavy loads over long distances without stopping for fuel.

The dream of commercial transatlantic flights across the Atlantic remained tantalizingly close to being realized. Although Pan American's Sikorsky S-42 flying boats had been built in 1934 especially for long-distance flying, it was becoming clear to the company – just three years later – that the planes were already obsolete. Rather than start commercial transatlantic flights with its Sikorskys, Pan American opted to wait for a new aircraft, which was being designed and built by Boeing. In 1938, Pan American suspended its transatlantic program while Boeing finished fabrication of the first of its new fleet of planes.

As for Imperial Airways, it also faced limitations to its flying boats; in order to fly non-stop across the Atlantic in 1937 the *Caledonia* had been stripped of all unnecessary weight and carried no cargo or passengers.

Initially, during the early 1930s, Imperial Airways believed it had a winner with its S-23 Empire Class flying boats, manufactured by Short Brothers. The planes were built to capitalize on a decision by the British government, in the mid-1930s, to transport all first-class mail by air. The S-23s allowed Imperial Airways to operate passenger and airmail flights to locations around the world. This made the flying boats a vital element in linking distant outposts of the British Empire.

The S-23s proved so successful that Imperial Airways ordered 42 of them. However, the planes could not match the range of Pan American's fleet of Sikorsky S-42s. But unlike their US rival, which, after the 1937 trial flights, focused on building a plane that could efficiently fly across the Atlantic, Imperial Airways opted to tinker with its S-23s.

In 1938, Imperial Airways tried an innovation called the "piggyback," which featured a Short S-23 flying boat carrying a smaller seaplane. It was dubbed a Short-Mayo Composite, in recognition of its designer, Robert Mayo. The concept was simple: the larger aircraft carried the smaller one part of the way across the Atlantic. Midway across, the seaplane was released to continue the journey to North America, while the transporter returned to base. The piggyback was designed solely to carry airmail across the Atlantic.

The "piggyback" innovation allowed Imperial Airways to fly airmail non-stop to Montreal from Ireland, as well as on other long routes. The *Maia* acted as a booster aircraft for the *Mercury* seaplane by flying it partway out to sea, and then releasing the smaller aircraft so that it could complete the journey.

The first transatlantic piggyback flight took place on July 20, 1938. Under the command of Arthur Wilcockson, a modified flying boat, named *Maia*, took off from Foynes. On its back the transporter carried a four-engine floatplane called *Mercury*. Wilcockson had flown to New York from Foynes on the inaugural transatlantic flight by the *Caledonia* a year earlier.

After several hours flying west over the Atlantic, the *Mercury* was released, whereupon Wilcockson turned the *Maia* around and headed back to Ireland. Because the *Mercury* had used little fuel – except during takeoff – to reach the mid-Atlantic, its pilot, Donald Bennett, was able to fly directly to Montreal with a cargo of 450 kilograms of mail without stopping first in either Botwood or Shediac. The entire journey, from Ireland to Montreal, took just over twenty hours to complete.

Although the piggyback experiment had proved successful, it was also cumbersome and expensive. The idea was abandoned after only one year.

By 1939, Pan American was ready to start commercial passenger and airmail service across both the North and South Atlantic. The airline introduced its new fleet of Boeing 314s, revolutionary aircraft in terms of flying range as well as passenger comfort and safety.

On June 24, 1939, a Boeing 314, named *Yankee Clipper*, took off from New York on the first airmail flight along Pan American's new northern route. Under the command of Harold Gray, it landed at seaplane bases in Shediac, Botwood, and Foynes, before finishing the journey in Southampton, England. Passenger flights commenced the following month. Weekly return flights continued until war broke out in September, whereupon Pan American ceased flying to Southampton for the remainder of 1939 and instead finished the flights in Foynes.

While Pan American was flying both passengers and airmail using the new, advanced Boeing aircraft, Imperial Airways was relegated to using its outdated planes. In order to carry mail across the Atlantic, modifications were made to allow for the mid-air refuelling of two S-23 flying boats. Handley Page Harrow air tankers were located at airports in Gander and Shannon, Ireland, to provide mid-air refuelling. Two flying boats, named *Caribou* and *Cabot*, made sixteen experimental transatlantic flights during August and September 1939 on Imperial Airways' New York-Montreal-Botwood-Foynes-Southampton route.

In the months prior to the outbreak of war in September 1939, two other airlines conducted transatlantic trial flights through Botwood. Air France completed three flights, but ceased further transatlantic activities following the declaration of war. American Export Airlines, a new US airline, made several trial flights using PBY Catalina twin-engine flying boats. American Export planes would become regular visitors to Botwood from 1942 to 1945.

All four airlines had been ready to offer passenger flights across the Atlantic when the Second World War intervened. Suddenly, military prerogatives took precedence over civilian priorities. The immediate effect on Botwood of the war in Europe was a reduction in air traffic; due to official US neutrality Pan American and American Export were forced to cancel their transatlantic flights through Shediac and Botwood after 1939.

The British also curtailed their transatlantic flights during the first two years of conflict. In 1940, British Overseas Airways Corporation (BOAC), which had been created with the merger of Imperial Airways and British Airways, made eleven Atlantic crossings. The planes, which carried mail and passengers, had been improved since the previous year and were now capable of flying across the Atlantic without mid-air refuelling. BOAC completed eleven further flights in 1941, after acquiring two of the superior Boeing 314 Clippers.

Botwood had become a busy community in 1941; the Royal Canadian Air Force began construction of a seaplane station and army base, thereby boosting the community's population dramatically. Britain's Royal Air Force, which had taken control of the civilian flying boat base, sent personnel to Botwood from the newly built Gander airport whenever a seaplane was scheduled to arrive.

The Botwood base proved useful to the military as well as to civilian airlines during the war. The RCAF conducted anti-submarine patrols from Botwood, mostly using Consolidated PBY Canso flying boats (a Canadian-built version of the Catalina flying boat). Many powerful political and military leaders passed through Botwood during the war, including US President Franklin Roosevelt, British Prime Minister Winston Churchill, and Lord Louis Mountbatten.

American Export resumed its transatlantic flying boat service in 1942. Under contract to the US Navy, the airline flew passengers across the Atlantic until the end of the war.

As for Pan American, it had maintained its southern transatlantic flying boat passenger service throughout the early years of conflict in Europe. Its flying boats flew between New York and Lisbon, Portugal, via Bermuda and the Azores. Pan American resumed passenger service on the northern passenger route to Ireland in 1943, several months after the US declared war on Germany.

Flying boats may have been high-class transportation, but travelling by air was still a relatively dangerous way to travel. Two flying boats were destroyed in accidents at Botwood during the Second World War, killing a total of 21 people.

On October 3, 1942, the *Excalibur*, a Sikorsky S-44A operated by American Export, crashed during takeoff, killing all fifteen passengers and crew. Levi Skinner, who was in charge of arrivals and departures at Botwood, recalled the rescue effort:

> I shot the boat up alongside the aircraft near the tail section that was afloat, and I could see a man's hand sticking out about three feet [one metre] under water; so I reached down and got a hold of his hand and squeezed it to see if there was any life in it, but there was no sign of life. I pulled on the hand to try and get the body aboard the boat, but the body was stuck in the fold of the plane where it had cracked off. I realized that if I just left the dead body there that it would probably be swept away by the current, so I tied a piece of rope around the wrist and secured it to the plane's fuselage.

A US Army ring, with the name Colonel Whitaker inscribed on the inside, had slid off the dead man's finger and into Skinner's hand. He contacted the family of the dead officer and returned the ring to them. Six more people died on November 8, 1943, when a RCAF Canso flying boat crashed into the Bay of Exploits while touching down at Botwood. There were five survivors.

Once the conflict in Europe ended in 1945, BOAC, Pan American, and American Export quickly abandoned Botwood in favour of Gander. Between 1937 and 1945 Pan American, the most active passenger carrier, had landed 455 times in Botwood, carrying 6,400 people across the Atlantic. In the decades following the Second World War the community of Botwood preserved the legacy of this brief, but important, era of passenger flight using flying boats. The slipway used by the planes has been maintained, as well as important artifacts, buildings, and wartime installations.

Land-based airports rendered obsolete almost all flying boats at war's end except in areas of the world where airports were scarce, such as Southeast Asia. The last regular passenger flights using flying boats were terminated in 1967. In addition, the armed forces of several nations, including Canada, continued to use Catalina and Canso flying boats for surveillance until the mid-1960s. Many of these planes were converted into water bombers for fighting forest fires, a role some of them maintained until the early 1990s.

CHAPTER 10

The Newfoundland community of Gander was built as an airport town. During the Second World War it served as a transit point for sending thousands of newly built planes to combat in Europe. A group of Hudson bombers is being prepared for their transatlantic journey.

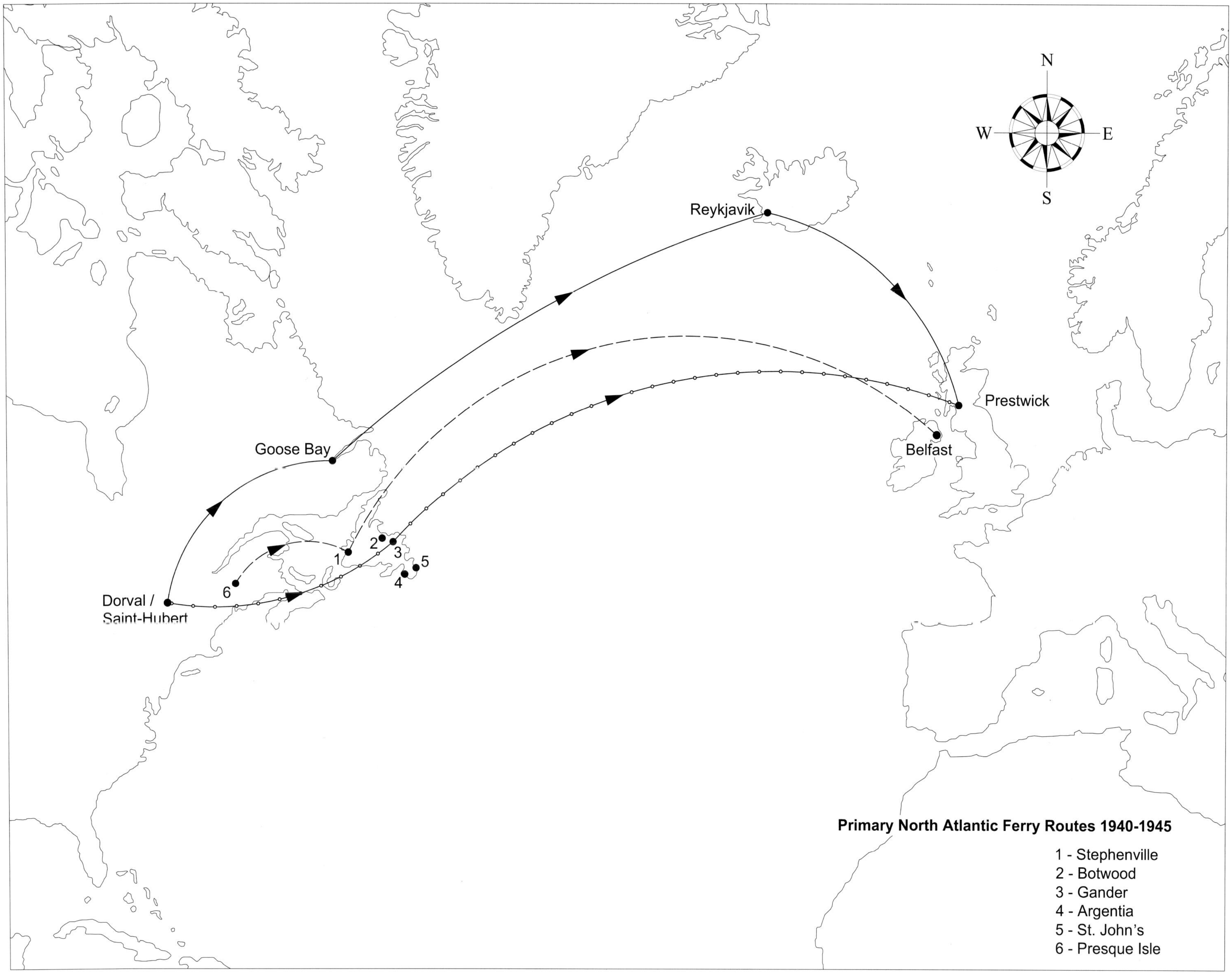
N
W
E
S
Reykjavik
Prestwick
Belfast
Goose Bay
1
2
3
4
5
6
Dorval /
Saint-Hubert
Primary North Atlantic Ferry Routes 1940-1945
1 - Stephenville
2 - Botwood
3 - Gander
4 - Argentia
5 - St. John's
6 - Presque Isle

The pioneer years of transatlantic aviation ended abruptly on September 3, 1939, when Great Britain declared war on Germany. Suddenly, long-distance transoceanic air travel had departed the realm of novelty and entered a new era of military necessity.

With German forces overwhelming much of Europe throughout 1939-1940, Britain girded itself against invasion by building aircraft capable of attacking enemy ships and troop carriers, as well as to counter the bombers that were pulverizing English cities and industrial infrastructure. In addition, Britain needed to bolster its own bomber squadrons in order to mount attacks on German rail lines and factories.

British industry could not build planes quickly enough to match the Germans; so the United States, although being officially neutral, began supplying Britain with aircraft – initially Lockheed-made Hudson bombers – soon after the war began. During the initial months, newly manufactured planes had to be towed over the Canadian border at North Dakota in order to ensure that the US adhered to its strict neutrality laws. The planes were then flown to eastern Canada and transported by ship across the Atlantic.

When hostilities began, British wartime planners had discounted the idea of flying bombers directly from North America, via Newfoundland, to Europe for several reasons: a lack of proper airport infrastructure in Newfoundland, a shortage of trained air navigators, the possibility of attacks on bombers as they were being flown overseas, and concerns about flying aircraft over the harsh North Atlantic – particularly during the notoriously treacherous winter months.

By the summer of 1940, manufacturers had orders for 26,000 aircraft. However, sending such a large numbers of planes by ship to Britain from Canada was difficult because of the toll exacted on merchant ships by German U-boats early in the war. Time was also an important consideration; the British were reeling from German success in continental Europe, and without a faster method of transporting new planes the English faced the possibility of having insufficient equipment to counter an invasion.

Therefore, in mid-1940 Britain's Minister of Aircraft Production, Canadian-born Lord Beaverbrook, decided to test the feasibility of transporting bombers to Britain from North America by air. To coordinate the experiment Lord Beaverbrook asked the Canadian Pacific Railway to handle the logistics.

The viability of flying newly built warplanes from North American factories to Great Britain was proven in November 1940, when seven Hudson bombers flew non-stop from Gander, Newfoundland, to Aldergrove, Ireland.

The operations room at the Newfoundland Airport in Gander, seen here on March 19, 1942, kept track of flights between England, Gander, and the Botwood seabase.

In November 1940, a group of seven Hudson bombers, led by Captain Donald Bennett of the Royal Air Force, left Saint-Hubert airport near Montreal and arrived at the recently built Newfoundland Airport adjacent Gander Lake. Known as Hattie's Camp until the mid-1930s, Gander had been recommended as the site for a transatlantic airfield in 1936 by Newfoundland aviator Douglas Fraser. The land-based airport had been built following the signing of the 1935 North Atlantic Civil Aviation Agreement. With this pact, Canada, Great Britain, Ireland, and the British-controlled administration in Newfoundland selected suitable locations for constructing new airports along the Atlantic coasts of North America and northern Europe that would be capable of handling flying boats and land-based transatlantic aircraft.

The Gander Lake area had several advantages over other locations in Newfoundland, outstripping even Harbour Grace, which at this time had become the favoured refuelling stop for transatlantic aviators. However, not only was Gander Lake free from coastal fog, it had been designated as an alternate landing site for flying boats in the late 1930s when sea conditions were poor at the flying boat terminal in Botwood. In addition, a rail line ran alongside the lake.

Most importantly, Gander Lake lies on a great circle route between North America and Europe. Great circle routes, which refer to the shortest distance by air between two locations on the globe, were particularly important at a time when the aircraft were largely unproven.

By 1940 Newfoundland Airport consisted of a single large hanger, an administration building, and a few support structures. On November 10 Bennett and the seven Hudson bombers took off from Gander on their experimental flight. Several hours later all of the aircraft landed safely at Aldergrove, Ireland. This achievement proved the viability of nonstop transatlantic flights, and Bennett's success would soon lead to the emergence of several military airports in eastern North America, including Dorval west of Montreal and several communities throughout Newfoundland and Labrador.

For the first six months after

Bennett's initial flight the CPR, in conjunction with the British Ministry of Aircraft Production, ran the ferrying of planes to Prestwick, Scotland. But the plan, which looked feasible on paper, proved much more difficult to carry out, and the entire operation soon came close to collapse. Although American factories were manufacturing planes in record numbers, neither the CPR nor the Ministry could find enough pilots and navigators to fly them. Dorval and Gander airports soon resembled massive parking lots, with planes filling any spare space available.

In May 1941 the Ministry cancelled its contract with the CPR and assumed complete control by coordinating the operation under a new body, the Atlantic Ferry Organization (ATFERO). But when the bottlenecks persisted, the British government shut down ATFERO and turned control over to the Royal Air Force, which created Ferry Command.

The RAF and the Royal Canadian Air Force solved its manpower problem by offering transatlantic flying experience to pilots, navigators, and wireless operators prior to their joining squadrons in Europe. In addition, many civilians who trained aircrews in Canada for the British Commonwealth Air Training Plan volunteered to ferry planes.

As the war went on, planes began carrying passengers and a growing volume of mail, cargo, and ammunition. To improve efficiency of handling the logistics, the RAF conducted a final reorganization in March 1943, with all ferrying functions grouped under a single umbrella. Ferry Command became Number 45 Group of the RAF's Transport Command, with headquarters remain-

A B24 bomber, which crashed on Christmas Day in 1943, was one of dozens that met a similar fate at Gander during the Second World War.

The Happy Valley-Goose Bay base became one of the world's busiest as a point of departure for planes heading to Europe. Housing hundreds of military personnel proved such a challenge that the Americans established a tent city for their men. Here, as at other military bases, soldiers fraternized easily with residents (above right). The US also established bases on the island of Newfoundland in Gander, St. John's, Stephenville, and Argentia (right, seen during the groundbreaking ceremony).

ing in Dorval.

Ferry Command would bring about a significant change to Gander's airport. During the early months of the war, the Royal Canadian Air Force had used the airport to conduct coastal surveillance and to accompany merchant supply convoys part way across the Atlantic. But with ATFERO and then Ferry Command, the airport required expansion, luring hundreds of construction workers to new jobs in central Newfoundland.

By mid-1941 Gander had been transformed into a massive military base, boasting three paved runways and living quarters for thousands of Canadian and British armed forces service personnel and civilians. Gander became an integral cog in the Allied war effort as a refuelling and maintenance base for flying newly manufactured bombers – including Hudsons, Liberators, and Flying Fortresses – to Britain. When the war ended in 1945, some 20,000 aircraft had stopped at Gander on their journey across the Atlantic.

The success of the Dorval-Gander-Prestwick route convinced the Allies to expand ferrying operations. They established two additional routes, one across the southern Atlantic, which linked the US to Egypt, using bases in the West Indies, South America, Ascension Island, and Africa.

A third route would go into the far north, using a network of bases in Labrador, Greenland, and Reykjavik, Iceland. This decision altered the lives of people living in one of the most isolated regions of North America. Until 1941, the aboriginal populations of central Labrador had had little contact with the outside world. This changed radically after the Canadian government agreed to build a military air base at the mouth of the Churchill River.

The Canadians completed the

Among those who offered support to US troops during the Second World War was Frank Sinatra (right), who visited Stephenville's Harmon base in 1945. Other dignitaries included Franklin Roosevelt and Winston Churchill (above); they discussed military strategy meetings on a British battleship near Argentia in Placentia Bay in August 1941.

Goose Bay base and its three runways in just a few months by bringing several thousand construction workers to central Labrador. During the war, Canada, Britain, and the US also established bases at Goose Bay.

By 1942, 1,700 armed services personnel were working at the base, and at its peak during the war the town's population exceeded 9,000 people. The village of Happy Valley, which was originally known by its more ironic title of Refugee Cove by the throngs of migrant workers, became home to the many civilians who worked at the base. By war's end 24,000 aircraft had passed through Goose Bay, making it the busiest airport in the world.

Other areas of Newfoundland and Labrador were also greatly affected by the war. In 1941 the Canadian military selected land on the outskirts of St. John's adjacent the town of Torbay to build an airbase. Local aviators had warned the Canadians that the hilltop region was unsuitable for an airport due to the prevalence of fog. Nevertheless, by year's end Royal Canadian Air Force Station Torbay had been cut out of the wilderness, and in place of bogland were runways, taxiways, aprons, hangars, and other facilities.

Several thousand RCAF personnel would be stationed at Torbay during the course of the war. Planes stationed at the airport conducted coastal surveillance and hunted for German submarines that were plying Atlantic convoy routes.

Although the Canadian armed forces established a significant presence in Newfoundland and Labrador, it was perhaps the Americans who made the greatest economic and cultural impact during the war and the initial years of the Cold War. These ties were forged in 1940 with the Destroyers for Bases agreement, in

An overhead view of St. John's airport during the Second World War, where Canada and the US flew reconnaissance flights to protect convoys and detect enemy submarines. Left: Some visitors to Harmon Field were offered promotional material, such as this booklet, upon their arrival.

which 50 First World War navy ships were given to Britain in return for the right to build bases in British possessions throughout the Caribbean and Newfoundland.

Although the US would not enter the war as a combatant until the Japanese raid on Pearl Harbor in December 1941, it recognized the strategic importance of Newfoundland in protecting shipping lanes and supporting the war effort in Europe. In addition, Canada and the US feared the consequences of Newfoundland falling into German hands, since the island could potentially be used to launch attacks along the eastern seaboard of North America.

The impact of the Americans on Newfoundland was unprecedented; thousands of armed forces personnel flooded to the island and Labrador, where they built and operated new bases at Argentia and Stephenville, and also established their own bases alongside the Canadians and British in Gander, Goose Bay, and St. John's. To coordinate the transfer of bombers to Europe, the US set up its own organization, the Air Transport Command.

During the war Gander's population expanded from a few hundred civilian and military personnel to 15,000 Canadians, Britons, Americans, and Newfoundlanders, making it the largest airport community in the world. Between 1941 and 1945 thousands of aircraft passed through the base, with planes sometimes taking off at one-minute intervals over a period of several hours.

As for Stephenville, in 1941 it was already a thriving community at the northeast end of St. George's Bay on the west coast of Newfoundland. Many of its 1,000 residents worked in the lumber industry, supplying wood to a pulp and paper mill in Corner Brook. Farming was also an important source of employment in the area.

The lives of residents changed dramatically with the Destroyers for Bases agreement because the Americans believed that Stephenville, with its strategic location and temperate weather conditions, would be an ideal location to build a forward operating base in the event fighter aircraft were needed to defend the Atlantic region from attack. This threat never materialized, and the new airport was used primarily as an emergency landing field until 1943 when it became a refuelling stop for transatlantic aircraft. The airport, named Harmon Field, became part of the US Air Transport Command.

Most residents of Stephenville were relocated to make way for the new airbase, and the population surged to a peak of 7,000 people as facilities were built during the war.

Like Stephenville, many residents of the fishing community of Argentia were moved from their homes to make way for the Americans. The town's location on the east side of Placentia Bay on the south coast of Newfoundland was ideal for the US Navy due to its deep, sheltered harbour and flat terrain.

During the war, the Argentia base consisted of the Naval Operating Base and the Naval Air Station. Three runways were built on the north side of the harbour, with the Fort McAndrew Army base located on the south side. The army's primary role was to protect US naval forces at Argentia.

After the war, Gander airport was immediately converted to civilian purposes. One of the first airlines to use the airport was Pan American, which flew Lockheed Constellation aircraft on its transatlantic passenger service.

Like other Newfoundland and Labrador communities hosting military installations, the US presence in Argentia provided a substantial stimulus to the local economy, as thousands of civilians found work constructing the base and then hundreds more were employed throughout its lengthy operating phase. Unlike Goose Bay, Gander, and Stephenville, Argentia was not used as a base to ferry aircraft to Europe during the war. Instead, US aircraft stationed there provided protection to convoys and hunted for German submarines in the northwest Atlantic. The US Army coordinated its air operations from St. John's, mainly offering communications and transportation services in support of a large army base at Fort Pepperrell.

Although the vast majority of planes that stopped at bases in Newfoundland and Labrador arrived safely in Europe, accidents inevitably claimed dozens of aviators and their machines. The best-known victim was Frederick Banting, co-discoverer of insulin, who died in February 1941 when the Hudson bomber in which he was flying as a passenger crashed at Musgrave Harbour shortly after takeoff from Gander en route to England.

When the war in Europe ended in May 1945, the need for ferrying large numbers of aircraft to Europe immediately disappeared. Within weeks Gander airport was converted to civilian needs and by year's end several airlines, including Pan American World Airways, Trans-Canada Airlines (later Air Canada), and British Overseas Airway Corporation (later British Airways), moved to Gander to establish transatlantic passenger air service.

One immediate casualty of peace was the flying boat station at Botwood, which was permanently closed. The era of flying boats had ended for transporting passengers across the Atlantic; they had been rendered obsolete by land-based aircraft and the construction of modern airports.

But any expectations of further base closures in Newfoundland and Labrador were eliminated by a new international reality, the Cold War. While the number of American and Canadian armed services personnel stationed at the bases rapidly declined from wartime levels, the standoff between the Soviet Union and western nations extended the operational life of many wartime facilities.

Over the ensuing five decades Newfoundland and Labrador's airport towns evolved as the Cold War gained strength and then waned. Air surveillance became a key element in deterring Soviet long-range aircraft from encroaching upon Canadian and American airspace. In 1950 the US placed its Newfoundland and Labrador bases within the Northeast Air Command

Although Gander has lost its importance as a refuelling stop for many airlines, it still has a crucial role in aviation. Most notably, when US airspace was closed in the wake of terrorist attacks on September 11, 2001, dozens of passenger planes were able to make landings at Gander's expansive airfield.

(NEAC), with headquarters located at Fort Pepperrell.

However, the NEAC was eliminated in 1957 and all bases under its jurisdiction were placed under the control of the US Strategic Air Command, a division within the Air Force that was responsible for protecting US territory from potential attacks by the Soviet Union. This reorganization resulted in the closure of Fort Pepperrell, and the last of the American forces departed St. John's in 1960.

Throughout the 1950s and 1960s rapid advancements in radar technology resulted in the construction of a network of aircraft and missile detection stations across northern Canada and the high Arctic, principally the Pinetree Line and the Distant Early Warning (DEW) Line. This meant that aircraft patrols lost their importance, and fighter planes stationed in Goose Bay and Stephenville were largely limited to intercepting Soviet aircraft that had already been detected by radar.

Stephenville was also responsible for supplying and servicing US military installations in the northeast Atlantic region, including bases in Greenland and Baffin Island. In addition, it was the first major overseas stop for American military aircraft flying across the North Atlantic between North America, Greenland, and Europe. In 1953 Harmon became an in-flight refuelling station for US bombers involved in monitoring and defending the Arctic.

The US military continued to upgrade its Stephenville base throughout the Cold War years of the 1950s. By the mid-1960s, however, the era of supersonic bombers had rendered Harmon obsolete, since these new aircraft were capable of flying for extended periods of time without refuelling. The base was closed in 1966, but commercial aviation, which began at Stephenville in 1957, has remained an important component of the town's economy.

After the war, Fort McAndrew in Argentia became a US Air Force base, which in 1955 was transferred to the Navy. In 1968 the US began phasing out its air operations at Argentia and by 1973 the airstrip was closed. The base, which was primarily used to detect submarines in the 1970s and 1980s, was shut down permanently in 1994, thereby ending the American military presence in Newfoundland and Labrador.

In Goose Bay the Canadian base was reduced to that of an Armed Forces Station in 1968, and three years later they and the British personnel moved to the American side of the station. The US closed its Goose Bay operations in 1976, although the base was subsequently used as a training facility for members of the North Atlantic Treaty Organization (NATO).

Low-level flight training, which the British had introduced in 1951, expanded in the 1980s as many European members of NATO, including Germany and Italy, began sending pilots to Goose Bay. With the dissolution of the Soviet Union in 1991, however, several nations abandoned Goose Bay, and its importance as a training facility gradually declined.

Canadian and American military aircraft still use Gander and St. John's as stopovers on transatlantic flights. But since the Second World War it has been commercial traffic, rather than military aircraft, that keeps the runways busy. By 1950 Gander was known as the "Crossroads of the World," with many international flights between North America and Europe stopping at the airport for fuel. During the 1950s and 1960s Gander became one of the world's busiest airports.

But Gander's importance began to wane as improvements in jet engines steadily increased the range of commercial aircraft. As a result, landings in Gander by major airlines began to decline during the 1970s. One of the last passenger airlines to visit Gander on a regular basis was Aeroflot, which used Gander International Airport as a stopover between Russia and Cuba. Following the collapse of communism in eastern Europe during the late 1980s and early 1990s, Gander became a significant migration point for asylum seekers who left Aeroflot and Cubana flights at the airport in order to claim refugee status in Canada.

But while Gander became a source of hope for refugees, it has also seen its share of despair. On September 18, 1946, a Belgian Sabena Airlines DC-4 passenger aircraft crashed into a bog near Gander, killing 28 of 44 people on board. In 1967, a Czechoslovakian airliner crashed on takeoff from Gander, killing 37 of the 69 passengers and crew. The worst aviation disaster on Canadian soil occurred on the outskirts of Gander International Airport in December 1985 when an Arrow Air DC-8 crashed shortly after takeoff. There were no survivors when the plane, which was carrying 248 members of the US Army 101st Airborne Division and eight crew to Fort Campbell, Kentucky, disintegrated and burned. The soldiers were flying home for Christmas after completing a peacekeeping mission in the Sinai Desert.

Although Gander's status has changed significantly, the airport retains its strategic importance in the world of aviation. This was illustrated following coordinated attacks on the United States by terrorists who, on September 11, 2001, hijacked four passenger aircraft, destroying the World Trade Center in New York and a section of the Pentagon in Washington DC. US authorities immediately closed American airspace, forcing 133 large aircraft to land at air-

St. John's International Airport is the largest airport in Newfoundland and Labrador. Among the airlines using the facility is Cargojet, which operates Boeing 727 aircraft (shown here in 2006).

ports in Atlantic Canada. Gander, a town of 10,000, was among the busiest, with 38 jumbo jets making emergency landings at its massive airfield. The people of central Newfoundland played host to 6,600 passengers and crew until US airspace was re-opened several days later.

In St. John's Canadian civilian authorities assumed control over the airport in 1946, as passenger and freight traffic became increasingly important. The military continued to play a role at the airport, as the Department of National Defence reactivated RCAF Station Torbay in 1953. Canada signed an agreement with the US that allowed the Americans to occupy several airport buildings and locate an Air Force squadron at the facility; this arrangement proved to be short-lived as the US pulled out of both Fort Pepperrell and Torbay in 1961. The RCAF turned the airport over to the federal Department of Transportation in 1964.

Unlike Gander, which experienced a decline in traffic due to technological improvements that extended the range of passenger aircraft, the frequency of domestic and international flights going through St. John's International Airport has continued to grow. In 2006, St. John's handled 1.25 million passengers on domestic and international flights.

EPILOGUE

Children play in the foreground of the Vickers Vimy bomber in St. John's, which was flown across the Atlantic Ocean in 1919 by John Alcock and Arthur Whitten Brown.

Today, flying across any of the world's oceans is largely taken for granted, as the jet age has turned air travel into a safe, convenient, and economical way to travel. The passage of time has left only fading memories of that brief era when pilots were heroic figures – when exploits in the air were closely followed by entire nations.

Air races, whether across continents or over oceans, ceased to attract much attention following the Second World War, as flight became a commonplace activity. By then, the competition between airships and fixed-wing aircraft had been settled, and newspapers were no longer offering cash prizes for being the first to fly between the world's great cities.

It is now difficult to comprehend the degree to which people became enthralled by pioneering pilots and their machines. Undoubtedly, this interest was fuelled by novelty, along with a desire to forget the carnage of the First World War and then the devastation wrought by the Great Depression.

But this is only part of the answer, for surely the men and women who took to the skies were answering the question people ask as they gaze in awe at birds soaring above them: "What would it be like to fly?"

IMAGE CREDITS

Key: B Bottom; C Centre; L Left; R Right; T Top

Abbreviations
AP – Associated Press; Byrd – Byrd Polar Research Centre; AMD – Archives and Manuscripts Division Queen Elizabeth II Library, Memorial University; MHA –Maritime History Archive, Memorial University; NYT – New York Times; PANL – The Rooms Provincial Archives Newfoundland and Labrador; SI – Smithsonian Institution.

Chapter 1
pp. 4 PANL A12-11; 5T PANL B21-161, 5BR PANL B21-158, 5BL AMD 232-3.04; 6L SI 90-2211; 7T AMD 232-3.05, 7B AMD 232-3.05; 8 Trepassey Historical Society; 9 US Naval Historical Center; 10 PANL B21-155; 11 Private Collection; 12T AMD 232-3.09, 12B PANL A47-15; 13L PANL A47-28, 13R PANL A47-22; 14-15 Library of Congress; 16 SI 2004-18936;17 SI-83-390.

Chapter 2
pp. 18 AMD 232-3.16; 20L Sheet Music: Private Collection; 20R PANL A46-119; 21T MHA, 21B SI 94-2; 22L Corbis, 22R PANL B21-177; 23 PANL B21-175; 24L Byrd 7744-5, 24R Byrd 7748-1; 25 Corbis; 26T SI 81-3420, 26B Private Collection; 27T Private Collection, 27B AMD 232-3.12; 28 AMD 232-3.14; 29 AMD 232-3.15; 30T AMD 232-13.18, 30B PANL A46-98; 31T PANL B21-80, 31B PANL B21-78; 32 AMD 232-3.19; 33 NYT 7 September 1927; 34 PANL A46-8; 35 PANL B21-46; 36 Byrd 7739-10; 37 Private Collection.

Chapter 3
pp. 38 PANL A17-88; 40L NYT 1 September 1927, 40R NYT 25 December 1927; 41 Corbis; 42L Corbis, 42R Central Press; 43 PANL B8-38; 44L *Anaconda Standard* 18 June 1928, 44R Private Collection; 45 Parson, B. 1983. *The Challenge of the Atlantic*. St. John's: Robinson-Blackmore Book Publishers; 46 SI 73-4032; 47 Trepassey Historical Society; 49 Corbis.

Chapter 4
pp. 50 Sheet Music: Jack Mills Music Publishers; 52L *Aviation Stories and Mechanics* June 1928; 52R Commemorative Coin: Private Collection; 53T *Toronto Daily Star*, 53B *Aviation Stories and Mechanics* June 1928; 54L PANL B21-72, 54R PANL B21-71; 55T AMD 232-3.29, 55B AMD 232-3.31; 56 AMD 232-3.34; 57 PANL H5-29; 58TL AMD 232-3.35; 58BL AMD 232-3.36; 58R PANL A46-125, 59 PANL B21-90; 61 Canadian Aviation Museum; 62 Canadian Aviation Museum.

Chapter 5
pp. 66 *Them Days* magazine; 67L www.pinetreeline.org, 67R SI 2004-22973; 68T SI 80-18087, 68B SI 80-18073; 69 *Chicago Tribune* 2 July 1929; 70 AMD 232-3.60; 71L *Illustrated London News* 11 June 1927, 71R *Fort Wayne Journal-Gazette* 2 March 1932; 72 Corbis; 73 Richard U. Light Foundation.

Chapter 6
pp. 74 PANL B21-18; 76 SI 00169975; 77 PANL B21-17; 78L *Evening Telegram* 15 July 1933, 78R Sketch Book Cover: Private Collection; 79T & B Private Collection; 81 PANL A57-12; 83 Postcard: Private Collection; 84 Corbis; 85 SI 73-8701.

Chapter 7
pp. 89 PANL A46-134; 90TR AMD 232-3.43, 90BL Fox Film Corporation, 90BR Private Collection, 90C Lapel Pin: Private Collection; 91T PANL A46-142, 91B AMD 232-3.44; 92 International News Photos; 93T AMD 232-3.52, 93C Wisconsin Aviation Hall of Fame, 93B PANL B21-65; 94L *Toronto Telegram* 23 August 1934, 94T Enna Jettick Shoes Card: Private Collection; 94B PANL A46-32; 95 PANL A46-22; 96T PANL B21-52, 96B Agence France Presse; 97L AMD 232-3.67, 97R AMD 232-3.69; 98 Corbis; 99 SI 2004 18938; 100L AMD 232-3.65, 100R Sheet Music: Irving Berlin Inc. Music Publishers; 101 PANL H5-20.

Chapter 8
pp. 102 AP; 104 AP; 105 AP; 106 Parsons; 107R AMD 232-3.44; 108L Postcard: Private Collection, 108R PANL B21-101; 109 SI 2004-18942; 110 AP; 112 *The Flying Family in Greenland*; 113 *The Flying Family in Greenland*; 115 *The Flying Family in Greenland*.

Chapter 9
pp. 116 AMD 232-3.72; 118 AP; 119 Corbis; 120 PANL B21-114; 121 PANL B21-107; 122 PANL A47-71; 123L PANL A46-153, 123R PANL B21-110; 124 PANL B21-171.

Chapter 10
pp. 126 PANL B21-183; 128 MHA PF 306-866; 129 PANL B21-200; 130-131 North Atlantic Aviation Museum; 132L MHA PF306-796, 132R Private Collection; 133 Private Collection; 134L MHA PF 306-407, 134R MHA PF 306-722; 135L MHA PF 306.407, 135R MHA PF 306.722; 136 North Atlantic Aviation Museum; 137 North Atlantic Aviation Museum; 139 St. John's International Airport Authority.

Epilogue
p. 139 Private Collection.

BIBLIOGRAPHY

Air Commodore H.G. Brackley: Air of Authority – A History of RAF Organization. <http://www.rafweb.org/Biographies/Brackley.htm>.

Allward, M. 1981. *An Illustrated History of Seaplanes and Flying Boats.* Ashbourne, United Kingdom: Dorset Press.

Almond, P. 1997. *Aviation: The Early Years.* Cologne, Germany: Konemann Verlagsgesellschaft.

Amelia Earhart. <http://www.acepilots.com/earhart.html>.

Archbold, R. 2005. Hindenburg: An Illustrated History. Toronto: Madison Press Books.

Aviation Stories & Mechanics (Summer 1928).

Balbo, I. 1934. *My Air Armada.* London: Hurst and Blackett.

Ball, N.R. (ed.) 1988. *Building Canada: A History of Public Works.* Toronto: University of Toronto Press.

Bonds, R. 2003. *A Century of Flight.* London: Salamander Books.

Botwood Heritage Society.

Bowman, B. and Parsons, B. 1983. *The Challenge of the Atlantic.* St. John's: Robinson-Blackmore.

Browne Papers. Queen Elizabeth II Library, Memorial University.

Byrd, R.E. 1928. *Skyward.* New York: G.P. Putnam's Sons.

Capelotti, P.J. 2002. *Technology and Memory at Port Burwell, 1929.* The Pine Tree Line. <http://www.pinetreeline.org/other/other8/other8cl.html>.

Cardoulis, J.N. 1990. *A Friendly Invasion: The American Military in Newfoundland 1940-1990.* St. John's: Breakwater.

Charles Nungesser. Legendary Aviators and Aircraft of World War One. 2003. <http://www.acepilots.com/wwi/fr_nungesser.html>.

Daily News. St. John's. September 1927.

Dunmore, S. 2004. *Undaunted.* Toronto: McClelland & Stewart.

Earhart, A. 1932. *The Fun of It: Random Records of My Own Flying and of Women in Aviation.* New York: Brewer, Warren & Putnam.

Elliott, J. May/June 2004. *Profile: Amelia Earhart.* Empowerment4Women.org. <http://www.empowerment4women.org/respect/mj04_ameliaearhart.html>.

Ellis, F.H. 1962. *Canada's Flying Heritage.* Toronto: University of Toronto Press.

Ellis, F.H. and Ellis, E. 1963. *Atlantic Air Conquest.* London: William Kimber and Company.

Amelia Earhart and the Friendship. Folklore of Trepassey. Trepassey, Newfoundland and Labrador: Trepassey Public Library.

Encyclopedia of Newfoundland and Labrador, Volume I (1981). St. John's. Newfoundland Book Publishers (1967) Ltd.

Ethell J.L. *Smithsonian Frontiers of Flight.* New York: Crown Publishing Group.

Evening Telegram. St. John's. September 1927.

Frances Wilson Grayson. <http://prion.bcls.uh.edu/george/grayson/fwgrayson.htm>.

Heppenheimer, T.A. *Hawker Siddely. Essays: Aerospace Industry.* U.S. Centennial of Flight Commission. 2003. <http://www.centennialofflight.gov/essay/Aerospace/Hawker/Aero51a.htm>.

Hutchinson, G.R. 1935. *The Flying Family in Greenland.* Binghamton, New York: Thomas Y. Crowell Company.

Hamlen, J. 1971. *Flight Fever.* New York: Doubleday & Company.

Hardesty, V. 2002. *Lindbergh: Flight's Enigmatic Hero.* Orlando, Florida: Tehabi Books.

Hawker Siddeley. <http://www.centennial of flight.gov>.

Holt, P. *R-34: An Atlantic Airship Exploit*. Aviation History, 2002. <http://history1900s.about.com/library/prm/blr34atlanticairship5.htm>.

Jablonski, E. 1972. *Atlantic Fever*. Toronto: Collier-Macmillan Canada.

Krippene, B.L. 1995. *Wingwalker: From Wisconsin to Norway, The Larson Brothers and Clyde Lee*. Friendship, Wisconsin: New Past Press Inc.

Lindbergh, A.M. 1938. *Listen! The Wind*. New York: Harcourt, Brace & World Inc.

Lindbergh, C.A. 1953. *The Spirit of St. Louis*. New York: Scribner's Sons.

Lost Adventurers: Daring But Not So Lucky. Pioneers of the Sky. (n.d.). <http://goodies.freeservers.com/lostadventurers.html>.

Luff, D. 1993. *Mollison: The Flying Scotsman*. Washington: Smithsonian Institution Press.

Mackworth-Praed, B. 1990. *Aviation: The Pioneer Years*. London: Studio Editions Ltd.

New York Times. 1927 to 1939.

Newsletter of the Inland Waterways of Ireland. Winter 2001.

Newfoundland Herald. January 2, 1988.

Newfoundland Quarterly. Autumn 1933.

O'Brien, P.J. 1935. *Will Rogers: Ambassaor of Good Will, Prince of Wit and Wisdom*. Philadelphia: John C. Winston Company.

Ohio State University Archives, Papers of Admiral Richard E. Byrd, RG 56.1, 7739-10.

O'Neill, P. 2003. *The Oldest City: The Story of St. John's, Newfoundland*. Portugal Cove-St. Philip's, Newfoundland and Labrador: Boulder Publications.

Post, Wiley. National Aviation Hall of Fame. <http://www.nationalaviation.org>.

R-34's Historic Round Trip. Gasbags. <http://www.airship.cwc.net/R34.htm>.

Record of Service of Wing Commander Charles Sandford Wynne-Eyton, DSO 186.

Rowe, P. 1977. *The Great Atlantic Air Race*. Toronto: McClelland and Stewart.

Royal Canadian Mounted Police Annual Reports. 1927 – 1929. Ottawa: National Archives of Canada.

Sikorsky, I.I. 1938. *The Story of the Winged-S*. New York: Dodd, Mead & Company.

Sinnott, S. *Street Names of Gander*. <http://www.k12.nf.ca/sptech/projects/Grassroots/Street-SephS/streetnames.htm>.

Smyth, R. 1997. *TheLindbergh of Canada: The Erroll Boyd Story*. Burnstown, Ontario: General Store Publishing House.

Them Days. September 1986.

Thomas, L. 1925. *The First World Flight*. New York: Houghton Mifflin Company.

Time. Volume 36, Number 1, 1927.

Toronto Daily Star. September 1927, April and June 1928, October 1929, August 1930, and July 1932.

Vance, J.F. 2002. *High Flight: Aviation and the Canadian Imagination*. Toronto: Penguin.

Ware, S. 1993. *Still Missing: Amelia Earhart and the Search for Modern Feminism*. New York: W.W. Norton and Company.

We Saw Him Land. Smithsonian. (May 2002): pp. 96-99.

Whitman, A. 1974. *Charles Lindbergh: An American Aviator*. <http://www.charleslindbergh.com/ny/100.asp>.

Wykes, A. 1967. *Air Atlantic: A History of Civil and Military Trans-Atlantic Flying*. London: Hamish Hamilton.

Zanetti, G. *Francesco de Pinedo in Canada – 1927*. <http://www.giorgiozanetti.ca/pinedo/depinedo.html>.

INDEX